P9-DGM-390

3 2510 07250 8914

12/98

Lucy Robbins Welles Library
95 Cedar Street
Newington, CT 06111-2645

COMETS

CREATORS AND DESTROYERS

DAVID H. LEVY

A TOUCHSTONE BOOK
Published by Simon & Schuster

⊼⊼

TOUCHSTONE
1230 Avenue of the Americas
New York, NY 10020

Copyright © 1998 by David Levy
All rights reserved,
including the right of reproduction
in whole or in part in any form.

TOUCHSTONE and colophon are registered trademarks
of Simon & Schuster Inc.

Designed by Irving Perkins Associates

Manufactured in the United States of America

10   9   8   7   6   5   4   3   2

Library of Congress Cataloging-in-Publication Data

Levy, David H., date.
    Comets: creators and destroyers/David H. Levy.
        p.   cm.
    Includes bibliographical references and index.
    1. Comets.   I. Title.
    QB721.L38   1998
    523.6—dc21   98-3113
                                                        CIP

ISBN 0-684-85255-1

ACKNOWLEDGMENTS

DURING THE TIME SET for writing this book, I had to undergo two major surgeries for cancer. Thanks to Drs. Sheldon Marks, Richard Lasky, and Stephen Webster, and my physician, Jeffrey Selwyn, I am now free of cancer and wish to praise their work. Because of their efforts, this book will not be my last.

Lisa Edwards and Arthur Klebanoff, my agents from Scott Meredith, worked very hard to help set the focus for this book and in its early editing phase. It was a pleasure to work with Elizabeth Bogner, my editor at Simon & Schuster.

Some material, especially from chapter 8, was adapted from my "Star Trails" column in *Sky and Telescope*, and I thank my friends at the magazine, especially Edwin Aguirre and Kelly Beatty, for their help.

As I completed each chapter, I asked Wendee Wallach-Levy and Dr. Art Boehm to read through it for clarity. Roy Bishop of Acadia University, Clark Chapman, and Bob and Valerie Goff added suggestions. Wendee and I also thank our daughter and son-in-law, Nanette and Mark Vigil, for their editing and title suggestions.

For Wendee—
my wife and my best friend
I love you

CONTENTS

GHOSTLY APPARITIONS IN THE NIGHT

I had watched a dozen comets, hitherto unknown, slowly creep across the sky as each one signed its sweeping flourish in the guest book of the Sun.

—LESLIE C. PELTIER, *STARLIGHT NIGHTS*, 1965[1]

COMETS ARE LIKE CATS: they have tails, and they do precisely what they want. This book tells the story of comets, from their origins at the start of our solar system to Comet Hale-Bopp, which so many of us recently saw. The book is designed to tell that story by bringing together, in a simple, brief, and interesting way, the diverse subjects of comets, life, and impacts. The story of comets is a story of beautiful ghostly apparitions. It is also a narrative of wanton destruction and of the dawn of life.

Comets also are messengers. One of the hottest science stories of the century began on the icy wastelands of Antarctica at the end of 1984, when a group of young scientists riding a snowmobile found a meteorite. That rock, it turned out, was an emissary from Mars that had come here to stay. Blasted out of the Martian rock by the crash of a comet or an asteroid perhaps a billion years ago, the rock traveled lazily about the Sun as ages passed before striking the Earth, bringing with it evidence that simple life forms might have once inhabited Mars. The tale of that rock and all its scientific implications brings us to ponder what else comets have brought us through the ages and ultimately leads to the story of the life and death cycles of planet Earth.

When Comet Hale-Bopp swung around the Sun in the spring of 1997, its bridal-veil appearance on the world stage attracted the attention of millions. One night, I walked outside a hotel to see groups of people huddled in the parking lot, pointing out the comet. Another evening, two somewhat inebriated people looked at me, then at the sky to point out "Comet Halley-Bobb." Whatever you call it, the Great Comet of 1997 put on a spectacular performance. *Newsweek*'s cover story pointed out that when it last appeared, around 2400 B.C., the pyramids of Egypt were relatively new. Although we do not know the comet's last orbit with such a degree of accuracy, the magazine's point was unerring: Hale-Bopp last viewed a very different picture of civilization on planet Earth.

The big comet's 1997 visit was an unsettled time on Earth, but for thirty-nine troubled souls it was deadly. On a cold spring morning, the members of San Diego's Heaven's Gate cult, believing that the comet offered them a journey to a new life, committed suicide. I was stunned when I heard this news. For me, comets are beautiful things, and searching for them is a happy task that I have done for more than thirty years. The idea that a comet would be associated with death does not make sense to me, but it did throughout most of the history of civilization. The famous illustration of a group of terrified soldiers gazing at Halley's comet just before the Battle of Hastings in 1066, preserved on the tapestry in a church in Bayeux, France, drives home humanity's perception that comets are omens of disaster. The word "disaster" was, in fact, made to show this role: it comes from the Greek "bad star."

Comets have had a bad rap for thousands of years, and it is only in the last two hundred years that science has been able to rehabilitate them. Instead of arriving on courses that would bring terrible calamity to some prince on Earth, they now circumnavigate the solar system following paths we call orbits. In 1979, one of the most astonishing stories in the history of science began as Walter and Luis Alvarez noticed, in a layer of Italian rock that dated back to the time of the elimination of the dinosaurs, the first evidence of cosmic impact. They proposed the revolutionary idea that a body from space collided with Earth, causing a mass extinction 65 million years ago. In 1991, a hundred-mile-wide impact crater was discovered, buried under Mexico's Yucatán Peninsula, that confirmed, or at least validated, their thesis. Was it possible that comets do not predict disaster, but that they actually cause disaster?

The discovery that three-quarters of the species of life on this planet, including the dinosaurs, might have been extinguished by a single strike from space brought home to the scientific community the idea that impacts played a role in the course of life on Earth. But an event was about to take place that would bring the new evidence of impacts to front pages all over the world. In the summer of 1994, a train of shattered remnants of Comet Shoemaker-Levy 9 slammed into Jupiter, the solar system's largest planet. It was the greatest explosion ever seen in our solar system. For several months after the collisions, dark clouds the size of Earth persisted over the impact sites. These strikes vividly displayed the awesome force that Nature can bring to bear in a comet crash. Suddenly, we had new ideas about our planetary heritage. The Earth has been struck by comets, and these collisions made drastic changes in the shape of life here.

There's more to this story. In 1986, a flotilla of space-craft found carbon, hydrogen, oxygen, and nitrogen in Halley's comet—the building blocks of life—in virtually identical amounts to their existence in humans. Is it likely that when comets struck the primordial Earth, they brought the materials that made life possible?

Comets: Creators and Destroyers tells the story of comets from this new perspective. Gossamer travelers that visit our sky from time to time, these flying balls of ice and dust have shaped the course of life on Earth. In a special sense, we are the children of comets.

CHAPTER ONE

FROM DUST TO DUST

—*I am like a slip of comet,*
Scarce worth discovery, in some corner seen
Bridging the slender difference of two stars,
Come out of space, or suddenly engender'd
By heady elements, for no man knows:
But when she sights the sun she grows and sizes
And spins her skirts out, while her central star
Shakes its cocooning mists; and so she comes
To fields of light; millions of travelling rays
Pierce her; she hangs upon the flame-cased sun,
And sucks the light as full as Gideon's fleece:
But then her tether calls her; she falls off,
And as she dwindles shreds her smock of gold
Amidst the sistering planets, till she comes
To single Saturn, last and solitary;
And then goes out into the cavernous dark.
So I go out: my little sweet is done:
I have drawn heat from this contagious sun:
To not ungentle death now forth I run.

—GERARD MANLEY HOPKINS, ENGLISH POET AND JESUIT PRIEST,
AFTER OBSERVING TEMPEL'S COMET IN 1864[1]

IMAGINE A PACKAGE OF rock and ice, a small world lost in the depths of space far beyond Pluto. No larger than a village, this world, which we call a comet, moves lazily around the Sun, which is so far away that it appears only as a bright star in its black sky. Suddenly, something— maybe a passing star—causes it to shift. Instead of moving around the Sun, the ball of rock and ice now moves in a new orbit that takes it closer to the solar system's central star. Over hundreds of thousands of years, the Sun brightens as the comet gets closer, passing Pluto and Neptune as it works its way inward. Jupiter's mighty clouds and storms beckon, but it ignores them in its relentless push toward warmth and the Sun.

Not long after it passes Jupiter, the Sun's rays start their slow work on the body's ices. One by one, they heat and turn directly into gases. After a period of quiet lasting the comet's entire lifetime of 4.5 billion years, the comet awakens, throws off its covers, and, like a peacock, spreads its material into space. Now surrounded by a cloud of gases and dust we call a coma, it forges ahead toward Mars, Earth, and Sun.

A BRAVE NEW WORLD

On Earth, it is May 20, 1990, and after giving a lecture about comet hunting to a group of visitors, I head home to my backyard observatory. Although the darkness of the predawn sky is lessened by a last quarter moon, I decide to begin searching for comets. Turning off white lights and

Tucson's two-hundred-year-old San Xavier mission must have seen several magnificent comets. Painting by James V. Scotti.

switching on red ones, so that my eyes will adapt to the darkness outside, I walk into the yard. There I have a garden shed that I had made into an observatory some years earlier.

Out in the garden shed observatory that morning, I grab a handle and give a shove. Slowly, the big roof starts moving, and a minute later, the observatory is open to the sky and I begin a slow search through the eastern sky sweep with Miranda, my sixteen-inch-diameter reflecting telescope named after the Shakespearean character who spoke of a "brave new world." Swinging the telescope north to keep as far from the Moon as possible, I "sweep" through the constellation of Andromeda, past the bright star Alpheratz, and on into the four-cornered figure of Pegasus, the winged horse.

Not far from that bright star, my telescope stops as a soft, fuzzy patch of light enters the field of view. I check several atlases of the sky, but none suggests that any fuzzy spot belongs in that lonely area. By now it is dawn, and my observing has ended. Twenty-four hours later, I look again. The "spot" is in a different position—it has moved! I have found a new comet—the holy grail of an observer. Now certain that the comet is a new one, I report it to the International Astronomical Union, which maintains a clearinghouse for discoveries called the Central Bureau for Astronomical Telegrams. After more than 4 billion years of anonymity, the new world is finally announced.

By early July of 1990, the new comet is experiencing the full warmth of the Sun. As it crosses the summer Milky Way, it grows a graceful tail as its ionized gas and dust particles escape.

GHOSTLY APPARITIONS

Comets first came to me one bright afternoon in the late 1950s. Just as our Hebrew school class was about to start, I lazily looked out the window and saw what seemed to be a comet hanging in the western sky, its bright tail calling our attention. The other children teased me as I wondered aloud how we would report this wondrous apparition. It couldn't have been a comet, though: the two major comets during the 1950s both appeared in 1957, and neither was bright enough to be seen clearly in daylight. My childhood comet was probably the vapor trail of an airplane, or a high cloud.

Had I brought the sighting to the attention of my

Hebrew teacher, he might have told us of a passage from the first book of Chronicles. It describes what could be a comet that appeared over Jerusalem during the time of King David. The biblical passage is read every year at the Passover seder:

> *And David lifted up his eyes, and saw the angel of the Lord standing between the earth and the heaven, having a drawn sword in his hand stretched out over Jerusalem.*[2]

Could a sword in the sky be a comet? We do know that the ancient Hebrews, like their Arabic neighbors, enjoyed looking at the night sky and sought meaning among its many stars and events. A bright comet, appearing once every two decades or so, would have attracted their attention as much then as now.

Despite the fact that I had never seen a comet, when I had to choose a topic for a sixth-grade public-speaking exercise in 1960, I chose comets. It was the first time I had ever given a speech in front of a group of people, and I was uneasy. I committed to memory every fact about comets that I could lay my hands on, but on speech day I was so nervous that I held a blank sheet of paper in front of me so I wouldn't have to look at my classmates.

The speech was a three-minute summary of what we understood about comets in 1960. Comets, I said, are large "dirty snowballs," balls of ices and dust, and they travel around the Sun in wide, looping paths, or orbits. The most famous, I said, was Halley's comet, which loops around the Sun every seventy-six years. Traveling all the way to a place beyond Neptune, the lonely comet then heads toward the Sun. Back in 1960, Halley's next visit in 1986 seemed a very long time in the future. I might have added that comets get discovered by people who, with

Comet Hale-Bopp, photographed by Jean Mueller on March 5, 1997. The observatory dome houses Palomar's forty-eight-inch-diameter Oschin Schmidt telescope.

small telescopes and perseverance, seek for them in the night sky.

The speech was pretty well received by my classmates. Sitting in the back, our teacher commented, "Nice speech, Levy. Can I see your notes?" Embarrassed, I looked down at the blank sheet of paper and the class broke out in laughter.

DIRTY SNOWBALLS IN THE NIGHT

Although I wasn't aware of it at the time, our understanding of comets had undergone a major change in the few years preceding my speech. Before 1950, comets were thought of as giant clumps of sand. That was the year that Fred Lawrence Whipple, an astronomer at Harvard, completed a study that seemed at first to be rather arcane. It concerned the little dust particles left in the wake of comets. We see those particles every night of the year; when they fall into our atmosphere, they heat the surrounding air until it glows. The result is a meteor.

For more than a century, astronomers have known that these meteor streams are the debris of comets. The dust particles are expelled from the comet as it rounds the Sun, and then travel independently around the Sun. Because they collide with each other, these particles do not stay in their cometary orbits for very long. Within a million years, most of them would spiral into the Sun.

The problem, Whipple perceived, is if comets were simply flying sandbanks, they would not be able to replenish the supply of meteors that had vanished as they fell into the Sun. Whipple proposed that a comet is a vast storehouse of ices, several miles in diameter, mixed with dust. When the comet is far from the Sun, the comet is inert. But as it nears the Sun and warms, the ices sublimate, or turn into gases, which leave the comet with explosive force. When this happens, particles of dust by the millions spread out into space.[3]

WHERE DO COMETS COME FROM?

For almost all of its life, a comet roams through the blackness of space far from Earth and Sun. One of the darkest masses of material possible, the black snowball is all but invisible. The solar system contains two major storehouses of comets. Both were proposed independently around the same time that Whipple was defining the structure and composition of comets. One lies in a belt beyond the orbit of Neptune. It includes the planet Pluto, and several observed comets of large size, more than one hundred miles across. The belt is named after Gerard Kuiper, a Dutch planetary scientist who spent most of his career at the University of Arizona in Tucson. The other repository, the ancestral home of Comet Levy, which I found in 1990, is an enormous sphere several trillion miles out. Called the Oort cloud, for Dutch astronomer Jan Oort, this sphere completely circumscribes the solar system.

The Oort cloud is the result of a long game of interplanetary pinball. The youthful solar system was filled with comets—in Earth's primordial sky there must have been dozens of bright comets at a time. Some of the comets collided with the planets. Others made close passes by planets, using their gravity to swing off into new orbits that would eventually land them near other planets. As the largest planet, Jupiter was the clear winner in this game. A comet swinging by Jupiter would get a gravitational hurl that would send the comet either out of the solar system forever, or off into the growing Oort cloud. Within about 500 million years of the solar system's birth, the pinball game was all but over, its supply of cometary materials exhausted. As we saw with the collision of Comet Shoemaker-Levy 9 with Jupiter in 1994, the process continues even today, but at a far more leisurely rate.

HOW FAST DO COMETS TRAVEL?

Most comets move much faster than the Earth's velocity around the Sun, which is about eighteen miles per second. Halley's comet makes a wide loop every seventy-six years. Its farthest point from the Sun is beyond Neptune.

When it is that far out, it parades through space very slowly; an airplane could probably keep up with it. As the comet moves in, it picks up speed. By the time it passes the Earth, it is sprinting along at close to forty miles per second.

HOW FAR DO COMETS GO?

To answer this question, we need to understand that comets, like people, cluster in families that are defined by their orbits. The best known is the Jupiter family, which consists of comets that have been so influenced by Jupiter's gravity that their orbits are related to that of the giant planet. In October 1990, Gene and Carolyn Shoemaker and I discovered an object that we thought was an asteroid. It looked like a faint star, and had no coma or tail. We observed it on several nights, reporting the positions to astronomer Brian Marsden in Cambridge, Massachusetts. When Marsden calculated an orbit for the asteroid he designated 1990 UL3, he was surprised to plot the object's moving from the vicinity of the Earth almost to the orbit of Jupiter, and back again in a period of several years. Its orbit was typical of a Jupiter-family comet, not an asteroid. He asked us to check the object's appearance. A careful look at the discovery films, taken through the eighteen-inch-diameter camera at Palomar Observatory in the mountains north of San Diego, still showed a star-like object. As a further check, astronomer Steve Larson and I pointed a much larger telescope—a sixty-one-inch-diameter reflector atop the Catalina Mountains north of Tucson—toward the object, then took a series of five-minute-long exposures. When we looked at the computerized results, near the bottom of the field dense with

Comet Hale-Bopp, photographed on April 7, 1997, using an F 3.5 135mm lens and equatorial tracking drive. Photo by Bob and Sue Liefeld.

stars was our object—with a faint tail forty thousand miles long!

The next day, the Central Bureau issued an announcement that asteroid 1990 UL3 was now Periodic Comet Shoemaker-Levy 2. The comet had given its identity away because its orbit was typical of a Jupiter-family comet. Comets in this family revolve about the Sun in periods averaging six or seven years. The fastest-orbiting comet is Encke, which races around in three and one-third years. Some Jupiter-family comets take much longer, heading out into the outer solar system. Halley's comet, the best-known member of the Jupiter family, has a seventy-six-year orbit.

Tethered to the Sun over periods of thousands of years are comets like Hale-Bopp, and the Great Comet of

1811. These comets travel incredible distances beyond the planets. The Comet of 1811 was visible to the naked eye for ten months. By December, the comet had a tail that covered almost a quarter of the sky. The comet was even credited with the coincidentally ultrafine wines from that year. I believe that the Comet of 1811 helped to inspire John Keats to compare the thrill of discovering a new work of literature to that of finding a new world. If this is true, then that long-departed comet, traveling to the very edge of the solar system and back in thousands of years, also traveled far enough to bridge the gap between science and poetry:

> Much have I traveled in the realms of gold,
>   And many goodly states and kingdoms seen;
>   Round many western islands have I been
> Which bards in fealty to Apollo hold.
> Oft of one wide expanse had I been told
>   That deep-brow'd Homer ruled as his demesne;
>   Yet did I never breathe its pure serene
> Till I heard Chapman speak out loud and bold:
> Then felt I like some watcher of the skies
>   When a new planet swims into his ken;
> Or like stout Cortez when with eagle eyes
>   He stared at the Pacific—and all his men
> Looked at each other with a wild surmise—
>   Silent, upon a peak in Darien.[4]

WHAT ARE ASTEROIDS?

Also called minor planets, asteroids are small, rocky bodies that orbit the Sun. The vast majority of the more than ten thousand known asteroids orbit the Sun between Mars and Jupiter, and are probably the remains of a planet near Jupiter that could never form because of the interference from Jupiter's gravity.

But not all asteroids orbit within the main belt between Mars and Jupiter. So many asteroids crowd the belt that there are inevitable collisions that send newly formed fragments on orbits that take them virtually anywhere. Some fifty thousand years ago, one of these asteroids, as big as a large room, slammed into an area now part of northern Arizona. This danger of asteroids and comets hitting Earth will be explored later in our story.

HOW DO ASTEROIDS DIFFER FROM COMETS?

From the point of view of an observer with a telescope, there is a simple way to tell the difference between an asteroid and a comet. Both types of worlds move among the starry background of the sky. As the Greek version of their name implies, asteroids look like starry points of light in a telescope. Comets, or long-haired stars, are moving spots of haze. Surrounded by hundreds of thousands of miles of escaping dust and gas, comets have a very distinctive appearance. If it's fuzzy and moving, it's a comet.

What happens when a comet is far from the Sun, and has no coma or tail? All that is left is the several-mile-wide "dirty snowball" of a nucleus. After a comet has made many trips around the Sun, it develops a crust. Eventually, the crust locks its volatile materials inside. Such comets look and behave like asteroids, and when someone discovers one, it is considered an asteroid until it displays the telltale coma or tail that defines a comet.

WHAT IS A METEOR?

Commonly described as a falling star, a meteor is not a star that has fallen out of the sky. After you see a meteor, there is not an empty space in the sky where a star used to be. A meteor is caused by a piece of dust, probably no larger than a grain of sand, that has entered the Earth's atmosphere. Friction with the air causes the grain to vaporize in a flash of light, and it is this light—not the tiny particle—that we see as a meteor.

Meteors are the rubble from comets, dust particles left off usually thousands of years before we see them appear like "shooting stars." As the Earth goes through its annual orbit of the Sun, it encounters streams of these particles. We call those encounters meteor showers. The most famous of these annual storms are the Perseids, which peak on August 12 each year. In early November, Earth cascades through a much weaker meteor stream called the Taurids. Twice each year, in May and October, we pass through the meteor stream from Halley's comet.

IS THE EARTH BEING PELTED BY A RAIN OF MICROCOMETS?

In the spring of 1997, a remarkable story hit the national news. NASA's POLAR satellite, launched to study the uppermost reaches of Earth's atmosphere, was recording hundreds of streaks that look like fiery trails of something entering the Earth's atmosphere. What could these streaks mean? If astronomer Lou Frank of the University

The sky is full of tiny meteoroids the size of grains of sand. In this photograph taken on July 21, 1962, Tim Hunter captured the American Echo 1 satellite. It was a monstrous balloon the size of a multistory building, and therefore a prime target for impacts by meteoroids. Notice how the streak of light, which shows Echo orbiting the Earth, is not even in brightness. The fainter parts of the trail occur because parts of the satellite have been deflated by meteoroid impacts.

of Iowa is correct, POLAR is recording a rain of micro-comets. His theory suggests that unlike village-sized regular comets, which are composed of ices and dust, these new objects, the size of mountain cabins, are made entirely of water ice.

Despite the heavy news play that this find generated, most astronomers doubt that the microcomets explain what the satellite observed. With so many of these things flying around, how come no one, from observant sky watchers on the ground to astronauts in space, has ever seen one? Even more important, if so many of these objects break up high above the Earth when they first encounter Earth's atmosphere, then many others should strike the Moon, which has no atmosphere. The Moon's surface should consequently be a skating rink paved with ice! Moreover, the seismometers left on the Moon from the Apollo years should have detected many impacts. The evidence, on both counts, does not support the idea.

If the microcomet theory does not hold water, so to speak, what is causing the images that POLAR sees? Can they be image defects, or leaks from the spacecraft? No one doubts that the craft is recording some phenomenon. But until more evidence is forthcoming, most scientists will consider the rain of microcomets more shimmer than shower.

WHAT DO COMETS HAVE TO DO WITH LIFE?

One of this book's major goals is to show the relationship between comets and the pageant of life on Earth. That relationship begins with an observation made of Halley's comet in 1986, when a spacecraft observed particles of carbon, hydrogen, oxygen, and nitrogen in almost identical proportions to their presence in us. C, H, O, and N, one scientist explains, make up the simple alphabet of life. It seems that the basis for life does exist in comets far more readily than it did in the early Earth, where temperatures were so high that any organic substances would have been vaporized. But temperatures never rose so high in the outer reaches of the solar system, where many comets waited out the formation of the planets. As the Earth cooled, comets delivered these substances as they collided with our planet.

There is a second side of the story of comets and life. The act of a collision between a comet and the Earth, as we shall see in chapter 5, is so violent that mass extinctions do result. Sixty-five million years ago, a comet collision set off an earthquake of magnitude 12 or greater, felt probably around the world. But what it set in motion was much worse. Secondary impacts of material wreaked havoc all over the world, an Earth-encompassing cloud sent temperatures plummeting, rain drenched with sulfuric acid pelted the ground: all this helped turn the Earth into a wasteland within a few weeks of the impact. It is not surprising that a mass extinction occurred.

HOW ARE COMETS DISCOVERED?

On November 14, 1680, Gottfried Kirch detected a new comet, becoming on that day the first person to discover a comet using a telescope. By the end of that year, the comet became bright enough to be seen at noon as it completed its hairpin turn around the Sun. Although other astronomers had found comets accidentally after that, it was Charles Messier's discovery of a comet eighty years later that was the first found as part of a deliberate search.

By 1770, having discovered several comets in this way, the French astronomer had become so famous that Louis XV called him the ferret of comets, and he was awarded a pension from his friend President Jean Baptiste de Saron of the Paris Parliament. De Saron was a man versed in comet orbit calculation, as well as in politics. But with the onset of the French Revolution, Messier was forced to leave the observatory in Paris. In the evening of September 27, 1793, Messier found a comet in Ophiuchus. As he had done so many times before, he informed his friend de Saron, who attempted to calculate an orbit using the positions Messier supplied. The comet was visible only briefly before it sank into the evening twilight.

However, by this time, de Saron was no longer president of the Paris Parliament. Accused as an enemy of reform, he was in prison awaiting execution. It is hard to imagine how de Saron could have cared about comet orbits when he was about to forfeit his head, but he did manage to calculate, from his prison cell, an orbit for Messier's comet. If de Saron's orbit was correct, the comet would move closer to the Sun, then swing away and reappear in the morning sky. On December 29, Messier searched the eastern sky and found his comet close to the position de Saron had predicted for it. Messier wrote of de Saron's last success and hid his note in a newspaper, which he was able to smuggle to the prisoner. On April 20, 1794, just three months before the end of Robespierre's Reign of Terror, de Saron was guillotined. Although Messier survived, his pension was gone, and the acclaimed astronomer was virtually penniless.

Messier's difficult life set an example for comet hunters to follow. In a sense, it is the world's slowest sport, in which scores are measured not in afternoons but in lifetimes. Messier's achievement of twelve comet finds has been the envy of searchers for the last two hundred years.

The key to a successful comet search is perseverance. Although some observers, like Alan Hale and Tom Bopp, do find comets by accident, most observers average some two hundred hours of search time for each comet they find. After I began searching in 1965, I spent more than 917 hours before I found my first comet nineteen years later in 1984.

Although comets may appear anywhere in the sky, the brightest ones usually are found within 90 degrees of the Sun—a quarter of the way around the sky. The evening sky in the west is often a productive area to search in the week after full Moon, as is the eastern sky in the morning around the time of new Moon.

It is important to know the sky well before starting a comet search, since the sky is full of fuzzy objects like galaxies and other objects that masquerade as comets. But it is even more important to note the difference between comets and galaxies, gas clouds called nebulae, and clusters of stars. Besides a noticeable difference in appearance, the most important thing is that a comet does move among the stars. The motion might be barely detectable over an hour or more, but comets do move.

HER TETHER CALLS HER

I discovered my fifth comet in May 1990. It grew brighter with each passing week, and, like a proud father, I watched it spread its wings across the sky all summer long. This comet was one of the highlights of my life. As the comet swung round the Sun that summer, it painted a beautiful picture in the night sky. But as it receded from the Sun, its tail shrank, and it faded slowly. Almost a full year after discovery, I said farewell to the fading comet now barely visible through my telescope.

Comet Levy is now far out in space, beyond the orbit of Saturn. It has virtually returned to its quiescent state, a ball of rock and ices, utterly frozen, the size of a village. It may never come around to our neighborhood again.

CHAPTER TWO

FOUR BILLION YEARS AGO

*This most excellent canopy, the air, look you, this brave
o'erhanging firmament, this majestical roof fretted with golden
fire. . . . What a piece of work is a man! How noble in reason!
how infinite in faculties! in form and moving, how express and
admirable! in action, how like an angel! in apprehension, how
like a god! the beauty of the world! the paragon of animals!
And yet, to me, what is this quintessence of dust?*

—SHAKESPEARE, *HAMLET*, 1604[1]

ONE OF THE MOST soaring tributes to humanity ever written in English, Shakespeare's lines are spoken by a prince in the midst of depression. What is this quintessence of dust? From the point of view of how the universe began and eventually led to humanity, this question is profound indeed.

In fact, were it not for a few conditions so obvious that we laugh at them, we could either not be here, or be much more comfortable! By examining these conditions, we get an idea of how fragile this o'erhanging firmament really is. I compile this list from thinkers like Nobel prize–winning physiologist George Wald, and from personal conversations with Clyde Tombaugh, who discovered the planet Pluto in 1930.

Life is possible on Earth because:

- The Earth orbits a small, single yellow dwarf star in the boondocks of our galaxy. But more than half of the stars in our galaxy are double or even triple suns. What if our home world circled one of these? Or what if we were a planet near the seething radiation at the center of the galaxy? In all likelihood, life as we know it would never have started.
- We have the right Sun. Astronomers call the star we circle a "G2" star. If our sun were a bit redder, like a G8, it would send out fewer ultraviolet rays, and we would not need an ozone layer or sun block to protect us. Life would be easier and more comfortable. But if the Sun were much different from what it is—say, a red giant, a blue supergiant, or a star that varies in brightness—then life as we know

it would not be possible. A giant sun would be so large that Earth would actually orbit beneath its surface, so that the planet would have to be located at a greater distance from the star. A blue star would send too much ultraviolet radiation. The changes in energy output of a variable star would make the planet's climate too extreme and unstable for life, as we understand it, to develop.

- Thanks to some cosmic impact that took place long ago, our Earth is tilted at an angle of 23.5 degrees. As the world revolves about the Sun, that angle means that for half the year, one hemisphere gets more direct sunlight. Changing seasons is the result. It would be nice if the tilt were less. If it were, the weather would be far more moderate all over, with fewer violent storms and a less extreme change in temperatures. But we're lucky the tilt isn't much greater; if it were, seasonal changes in weather would be devastating.

- Ice floats. If ice did not float, the ponds that graced the early Earth would have frozen solid, killing the delicate life within them and preventing higher forms of life from evolving. Instead, a floating roof of ice protected these ponds and the precious life they supported.

- The night sky is dark. The understanding of this simple idea began with Heinrich Olbers, a remarkable early-nineteenth-century German physician and amateur astronomer. He was a member of a planet-searching group dubbed the Celestial Police and the discoverer of the second asteroid, Pallas, around 1801. Like most of us, Olbers was amazed at the darkness of the night sky, but he also believed that the fact of darkness was a paradox: With so

Formation of the Moon: one year after impact. This James V. Scotti painting, specially commissioned for this book, is a representation of what the Moon, Earth, and sky looked like a year after the impact that formed it. The artist writes, "Here I imagine what it might have looked like about a year after that impact with the moon grown to nearly its full size. A disk of debris is still present, and a number of substantial-sized bodies still orbit. Several comets are visible in the sky and the Earth's surface is still quite molten and active. A large nebula hangs in the sky from which the proto-solar nebula emerged about 10 million years before this event."

many stars and other things in the universe, why didn't all their light come charging in on us, their radiation blasting us to death? Olbers never solved his riddle. He never knew about the nature of the universe he lived in. The preponderance of evidence indicates that it was created by a colossal Big Bang that started an expansion that continues, unabated, to this day. Because the universe is expanding, this radiation does not hit us all at once. Isn't it interesting to think that in the instant of primordial explosion that started the universe, the conditions for life on our planet far in the future might have been set?

AN EXPANDING UNIVERSE

How did we come to realize that the universe was expanding? Before an astronomer in northern Arizona did a small experiment, we had no idea how vast the universe really is. Just after the turn of the century, Vesto M. Slipher, an astronomer at the Lowell Observatory in Flagstaff, Arizona, photographed spiral-shaped clouds in space under the light of a spectroscope. It was a challenging project. Using the observatory's twenty-four-inch-diameter refractor, Slipher spent two full nights gathering enough light for a spectrum of a single small deep space cloud we call a nebula. Slipher was baffled by the results: all these spectra showed large shifts of light toward the red end of the spectrum.

Slipher was unable to interpret these results. He went on to do other things in his capacity as director of Lowell Observatory, including setting up the program that led to the discovery of Pluto in 1930. The red shifts would remain a mystery for some years afterward.

While Slipher had a twenty-four-inch-diameter re-

The Moon at crescent phase, with the dark part visible because of reflected light from Earth called earthshine. Photo by Steve Edberg.

fractor, Edwin Hubble had the great one hundred-inch reflector at Mount Wilson, east of Pasadena, California, at his disposal. He and Milton Humason took such clear photographs of nearby "spiral nebulae" that by 1924, they knew that these objects were not clouds at all but enormous groups of stars. No longer thought to be new solar systems, they were found to be galaxies like our own Milky Way. Hubble solved the mystery of the red shifts in 1929, by showing that they were the optical equivalent of a pitch of a train whistle as the train passes you. As the train approaches, its whistle stays at a steady pitch, but the moment the train passes, the pitch of its whistle drops. The faster the train is going, the more the pitch changes. This change in pitch, called the Doppler effect, is similar to the shift of light to the red end of the spectrum in a galaxy. The farther away a galaxy is, the faster it is racing away, and the more pronounced the red shift is in its spectrum.

Hubble's explanation, which most scientists accept to this day, is that the great superclusters are receding from each other because the universe itself, which began in a colossal explosion, is expanding.

In 1965, two scientists working for Bell Labs, Arno Penzias and Robert Wilson, set up a small antenna. They were puzzled by its detection of a steady radiation. After pondering this for some time, they came to two possible conclusions. The "radiation" was either a false note from the metal in a chicken coop on a neighboring farm (it wasn't), or it was the long-sought evidence of the background radiation left over from the primordial explosion

The lunar crater Copernicus was carved out of the Moon by a comet (or, less likely, an asteroid) impact. Photograph by Steve Edberg.

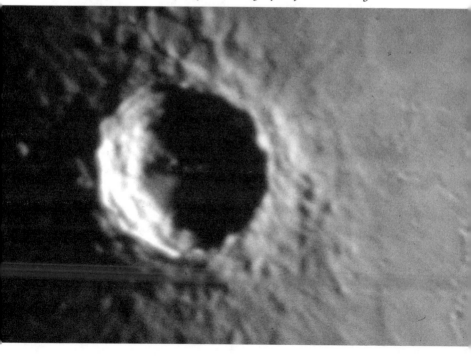

that began the universe. In 1989, the *Cosmic Background Explorer* satellite began its work on microwave radiation. The satellite succeeded in detecting weak signals that mapped the background radiation of the universe.

Several billion years might have elapsed between the instant of the Big Bang and the second instant in which our solar system, and the life that thrives on Earth today, were ready to form. During those billions of years, the galaxies, with their hundreds of billions of mighty suns, began to take shape and started spinning their mighty revolutions through space. Our own galaxy, which we call the Milky Way, was one of these; its big spiral arms rotate once in 225 million years. Four and a half billion years ago, our galaxy was a more dramatic place than now; it consisted of many big stars we call supergiants, stars with short lives that end with violent explosions. Scattered around these stars were large, dark clouds.

THE BIRTH OF THE SOLAR SYSTEM

Picture one of these large, dark clouds in space, sitting passively for an incredibly long time. Hundreds of light-years across, this "giant molecular cloud" was one of the biggest objects in the galaxy. Somewhere nearby, a massive star suddenly ran out of material for nuclear fusion. In a fraction of a second, the huge sun collapsed on itself, and then it blew itself apart in the awesome explosion of a supernova, spreading carbon into nearby space and into the large, dark cloud.

The explosion of a supernova is an incredibly violent event. We have seen only three in our own galaxy, in 1054, 1572, and 1604. Bright as the planet Venus, these stars seemed to appear out of nowhere. Viewed from outside the galaxy, they would have been as bright as the rest of the Milky Way galaxy put together. At their brightest, these exploding stars could be seen in daylight.

Almost 5 billion years ago, a supernova changed the nature of our giant molecular cloud. To the hydrogen and helium that made up the cloud were added organic material from the supernova, like carbon. A fragment of the cloud began to shrink, its tiny grains no more than a thousandth of a millimeter across, covered by a thin layer of organic molecules like frosting on a cake. Ancestral comets were formed by these grains slowly clumping together as the cloud started to rotate slowly. As the material in the cloud was whipped about as if it were in a gigantic, low-speed blender, particles joined together to form ancestral comets and some meteorites.

Some of these particles survived the eons of time to land intact on Earth as parts of meteorites we call carbonaceous chondrites. They have been found to contain grains of aluminum oxide, silicon carbide, and other organic compounds that were formed before the Sun. These microscopic grains are from stars now long gone, stars once spread throughout our galaxy.[2] It is likely that these grains, transported here by comets and carbonaceous meteorites, exist in some form in every living thing. We are made of ancient stars.

A lot has changed in the past 4.6 billion years. The organic grains in the large, dark cloud, which we call the solar nebula, were almost pure carbon, with some carbon monoxide (CO), water (H_2O), and formaldehyde (H_2CO). These are important materials in the chemistry in which simple organic materials combine to form biological molecules; for example, formaldehyde reacted to form carbohydrate sugars. These "prebiotic" reactions could have taken place in the original large molecular cloud, or later in the nebula that condensed to form the solar system, or later still in the solid clumps of material we call comets.

Comets by the trillions grew out of masses of grains that coalesced into solid clumps of material some one to ten kilometers in diameter. As the clumps ran into each other, they became larger bodies called planetesimals. It is also likely that as some planetesimals grow bigger, many collisions from smaller objects broke them apart.

Near the cloud's center, particles gathered more quickly, and temperatures started to rise. Smaller clumps surrounding the center, or protosun, become the nine planets. The trillions of comets were flying through the system, colliding with each other and with the larger clumps growing to become the nine planets of our solar system.

Probably not more than 100 million years passed between the initial collapse of the cloud and the crowning event of the ignition of the Sun. In the billions of years of galactic history, 100 million years is not a long time. But after the Sun's birth, events happened even faster. The planets grew to their present sizes quickly, and much of the cometary rubble was flung out of the inner solar system, either to the Kuiper belt beyond Neptune, or to the Oort cloud, or out of the solar system altogether.

A HOT TIME . . .

Once the Sun ignited, the solar system had the same basic shape and population as it does today. There were four small planets near the Sun, four very large planets farther out, and Pluto and her host of large comets. There were also a dozen or so other worlds, each having at least the mass of Mars. One of these Mars-sized "planetesimals" must have had an orbit more like a Jupiter-family comet than a planet. Swinging out toward Jupiter, then in toward the Sun, it passed by each of the inner planets, probably making several near misses over millions of years.

All of the worlds were very hot in those early years. The Earth's surface was probably hot enough to melt lead, until slowly, it cooled and formed a crust. Life on the primordial Earth would have been no picnic. Even on the coolest parts of the crust there would have been a constant threat of a cataclysmic volcanic eruption, along with a good view of several nearby volcanoes spewing forth lava and ashes. The sky would be dense both with stars and with comets—a dozen bright comets would be shining, their tails all pointing away from the Sun like an army in formation. The carbon dioxide–rich atmosphere had none of the life-sustaining oxygen we have today. Moreover, the Earth's sky had no moon.

THE BIRTH OF THE MOON

Every six or seven years in the sky over the primordial Earth, the Mars-sized planetesimal would make a close approach. Getting brighter by the night, it would shine prominently in Earth's sky for a few weeks, then back away toward the outer part of the solar system. Quite likely, this planetesimal had many dangerously close shaves with the Earth, as well as with Mars and Venus. Had a civilization been living on the Earth at that time, its scientists would have known that their species was living on borrowed time, and that sooner or later, there would be a collision.

When the titanic sideswipe finally happened, the

blow was inconceivable. According to the Harvard University team that modeled this event, the force of the collision was so strong that the Earth's crust melted away. The planetesimal disintegrated completely, its pieces roaring back up into the sky along with sizable chunks of Earth. A gigantic ring of particles—much larger and thicker than Saturn's rings—formed around the Earth. Imagine being able to see an episode like this!

In 1997, Shigeru Ida led a University of Colorado team in a careful experiment to learn how the ring of debris particles would coalesce into what we now know as the Moon. Ranging in size from a particle of dust to big bodies sixty miles across, these particles would have formed the ring at a distance of some fourteen thousand miles from the Earth. In all of the computer simulations, the particles clumped together within the space of a year! Additionally, more than half the particles in the ring eventually fell back to Earth instead of participating in the formation of the Moon.

The Colorado simulations produced a final interesting thought: in a third of the computer models, the debris formed two moons, not one![3] The two bodies would have orbited the Earth together, using each other's gravity to define unusual paths that would have resembled horseshoes around the Earth. In this scenario, one of the moons broke apart as it got too close to the Earth. Another possibility is that the two moons collided, forming the Moon we are familiar with today, and in the process caused even more debris to rain down on the hapless Earth.[4]

The violence of our Moon's birth was unusual, apparently, even by solar system standards. If these ideas are correct, perhaps we can look at the Moon with a different perspective when we recall this ancient nursery rhyme:

Lady Moon, Lady Moon, where are you roving?
Over the sea, over the sea.
Lady Moon, Lady Moon, whom are you loving?
All that love me, all that love me.

OTHER GREAT COLLISIONS

What happened to the other rogue planetesimals that cruised carelessly throughout the young solar system? One large world the size of Earth might have struck Uranus. The force of that impact caused the green giant planet to tilt over on its axis. The result is that Uranus rotates on its side, tilted at an angle of 98 degrees to the plane of its orbit. This results in some strange effects. During much of its eighty-four-year orbit around the Sun, either one or the other of its poles periodically faces the Sun and becomes the warmest place on the whole planet for up to half the planet's eighty-four-year revolution around the Sun.

Venus, Mars, Saturn, and Pluto, and, of course, the Earth, all rotate on tilts that would not have occurred just by the standard process of planet building. It is possible that each of these worlds was struck by one or more planetesimal bodies, all hurtling through the solar system along orbits that crossed the paths of the major planets. It is also possible that the same intruder that built the Moon might have torqued the Earth's axis of rotation to its present tilt of 23.5 degrees.

A COOLING-OFF PERIOD

The initial wave of megaimpacts passed gradually. Scientists are not certain which of two scenarios actually took place during the half billion years after the formation of the Moon and the departure of most of the large errant planetesimals. In one view, the solar system was filled with cometary bodies, each several miles in diameter, each a building block for the solar system, each a threat to a larger world. When the planets were small, collisions were frequent and of relatively low speed. But as the planets and the Sun grew larger, their gravitational pulls increased, and cometary collisions became more devastating. During the half-billion-year period after the ignition of the Sun, comets collided with the planets at an astounding rate—Earth was clobbered with a major comet hit perhaps every century—until some 3.9 billion years ago, when the period of heavy bombardment ended.

There is some evidence for a different description of the course of history. In this scenario, the initial bombardment faded out within several million years to a period of relative calm, when all but one of the rogue planetesimals had either collided or been expelled from the solar system. This last of the planetesimals worked its way through the realm of the inner solar system until a near miss with one of the planets caused it to break apart. The planetesimal's wreck resulted in near-total chaos as thousands of fragments slammed into Mercury, Venus, Earth, the Moon, and Mars. If this theory of a "late heavy bombardment" 3.9 billion years ago is true, then many of the craters we see on the Moon, and on Mercury, were formed during this period.

Some of the impacts of this time were so large they almost broke the Moon apart. When the Moon is near its full phase, a casual glance shows several round, dark areas that form the features of the legendary face of the "man in the Moon." Two of these—the eyes—are great impact basins, formed some 3.9 billion years ago during this violent period by the collisions of comets perhaps fifteen miles wide. Other large basins were instantly gouged out during this same time. The dark shading that identifies them so clearly is lava that came much later as the Moon resurfaced these basins.

It seems incredibly simple to visualize this history by studying the Moon through a small telescope. Most of the lava-filled impact basins occupy one large part of the Moon's visible surface. The brighter regions are mountainous highlands. The highlands, the oldest lunar surface, still show craters dating back to the earliest times of the Moon's history. However, the lava-filled impact basins are far smoother, with relatively few craters. This evidence indicates that the impact rate dropped sharply after the period of late heavy bombardment ended.

Although the rate slowed, it didn't drop to zero. Plastered over the Moon's surface are impact craters that are the centers of big systems of long, bright features called rays. They are composed of material forced out of the Moon by the shock of impact, and which then fell back to the Moon. The Moon's largest ray system belongs to Tycho, one of the largest craters. It was formed by the impact of what was most likely a comet, some 100 million years ago. In the course of the long history of the solar system, 100 million years is not a very long time. Visible through even a pair of binoculars, Tycho is evidence that the impact hazard, while low at this time, has not vanished.

THE MOON AND THE EARTH

Although the Moon formed only a few thousand miles from Earth, it obviously did not stay that close. Over billions of years it has moved farther from the Earth, and continues to recede at the leisurely rate of about three feet per century. Its orbital period of Earth is twenty-eight days, and forms the basis of several calendars, including the Hebrew and Arabic. The Moon and the Earth have been playing a game of tidal tug-of-war for the last 4.5 billion years. In this war of the worlds, the winner would slow down the other body's rotation period, and the Earth has clearly won. The Moon rotates in exactly the same period as it revolves about the Earth, a situation we call rotational lock. However, the Moon's weaker gravity does affect the Earth, slowing its rotation period down to its present twenty-four hours, thanks to a constant pull by the Moon. The rate continues to slow, and possibly billions of years in the future, Earth will also achieve rotational lock, its day a restful twenty-eight times longer than the present day.

In the meantime, sailors the world over have to plan their departures by "time and tide." Twice every day, ocean tides ebb and flow around the world, mostly according to the rule of the Moon. At some spots, the tides are amplified by a resonance factor. Like a child pounding his fists in the bathtub in time with the wave so that water eventually sloshes over the top, tidal resonance in the Minas Basin, in the Canadian province of Nova Scotia, causes the difference between low and high tide to top fifty feet. As the incoming tide pours water into the rivers off the basin, the inrushing water roars through the channels, filling them rapidly.

I had a chance to watch this tidal roar at the entrance to the Minas Basin. Looking down from a small grassy area at the edge of a several-hundred-foot precipice of basalt, I saw torrents of water roaring past. The incoming tidal flow was at its maximum, and as we stared down, a flow of water equal to all the rivers on Earth rushed by. It was hard to believe that this water was rushing with such speed, mostly in response to the gravity of the Moon.

ATTEMPTS AT LIFE

The period of late heavy bombardment was one of the most violent periods in the long geologic history of the Earth. Major impacts occurred on Earth and Moon every century. As I have mentioned, several of the last catastrophic impacts almost split the Moon apart as they formed the great impact basins. It was a time of utter devastation.

It was also the time when the first stirrings of life were delivered to planet Earth.

Too hot to support any organic materials, the primordial Earth was a place without any sign of life. And were it not for comets, the Earth would to this day be a place without organic compounds, while the outer solar system would continue to house them. Comets delivered the makings of life to the Earth. Over a long period of time, comets provided the Earth with its supply of carbon and its supply of water. In 1982, J. Mayo Greenberg, a theoretical physicist at the University of Leiden, suggested that a typical comet is composed of organic materials as well as water ice.[5] In March 1986, Greenberg's idea was proved dramatically by an actual visit to Halley's comet by the *Giotto* spacecraft, launched by the European Space Agency and named for the fourteenth-century Italian painter whose *Adoration of the Magi* might have been based on the artist's observation of Halley's comet. The spacecraft detected hydrogen cyanide, water, and formaldehyde as it passed by the famous comet.

These observations, supported with others from the ground, suggested that Halley is surrounded by very fine organic particles, now called CHON grains because they include carbon, hydrogen, oxygen, and nitrogen. These grains may be the seeds of life, as they contain the elements from which amino acids and other compounds essential to life are made.

In all likelihood, the long march of life on Earth began in a hail of cometary bullets. The dawn of life was a long period when comets and protoplanets were crashing into every planet. The Earth was a barren rock, its sky filled with comets, and impacts were happening frequently. In this hostile environment, life's building blocks were brought on the wings of comets, then crushed under the tremendous force of another comet strike, then brought again. In the violent atmosphere of the young Earth, life as we now perceive it was a far-off event indeed.

CHAPTER THREE

COMETS AND THE ORIGIN OF LIFE

ARE PEOPLE MADE OF THE SAME THINGS AS COMETS?

Atomic Percentage	Carbon	Hydrogen	Oxygen	Nitrogen	Total
In humans	9.5	63	26	1	99.5
In Halley	11.0	55	28	2	96.0[1]

THIS CHAPTER BEGINS NOT with a poem but with a table. Its simple numbers are poetry enough, for they compare the percentage of atoms of four elements in humans with Halley's comet. A poet might even suggest that if we give Halley's comet a brain and a human shape, then bring it to life somehow, it might as well play baseball or run for president! Other than the size of the cometary creature, its composition would pass for human.

This remarkable similarity is enough to suggest that we should look for a connection between comets and life. It is not the proof of a connection, but it is a start.

THE GREAT BOMBARDMENT

Pelted by a steady rain of comets, the Earth received a vast amount of water in the first few hundred million years after it was formed. On a young planet without the high mountains and deep oceans of today, comets dumped enough water to cover the surface of the planet to a depth of more than twenty feet! Some research suggests that comets contributed as much as ten times the present mass of water in all Earth's oceans.[2] As we know from chapter 2, a good amount of this water would have been vaporized by the impacts themselves, but the supply seems to have been more than enough.

If comets were the major source of water, they were not the only fountain of water in the primordial Earth. Lava, which contains small amounts of water, was carried by volcanic eruptions from beneath the surface of the Earth in quantities far greater than at present. Years ago, it was widely thought that volcanic events carried all of Earth's water, though we now know that volcanic material does not contain nearly enough to accomplish this crucial task.

DID COMETS BRING THE RAW MATERIALS OF LIFE?

The answer to this utterly profound question tells us something about who we are and the process by which we came to be. Comets formed slowly, as grain after grain clumped together. Some of these particles were organic, made up of materials from long-gone exploding suns. These grains enjoyed a lonely existence in space, meandering throughout the galaxy. They got irradiated by cosmic radiation, they collided with other grains, and eventually they massed together to become comets. A comet that shines in the evening sky might consist of grains from anywhere in our galaxy, formed at any time in the past. A grain with carbon formed in the awesome explosion of a supernova 5 billion years ago, another grain irradiated from a red giant sun someplace else—of such diverse particles are comets made. Any comet visiting us now carries with it the signature of our whole galaxy.

Comets brought these materials not as a gift of life, but as a gift of the raw ingredients that could become life. But they had to deliver them to Earth at the right time, or they could not have been the cause of life. Were comets needed in the first place? If the primordial Earth contained water, methane, and ammonia, then we do not need to invoke an exogenic source for the organic seeds. After all, the giant planets Jupiter and Saturn did, and still do, have such atmospheres. However, research within the

Comet Hyakutake passes to the lower left of the pole star in March 1997. Photo by David H. Levy.

past decade shows that the Earth formed at a far higher temperature than previously thought, resulting in a loss of virtually all of the primordial organic material. The consensus of geological thought reasons that by the time the planet began to cool, its atmosphere was mostly carbon dioxide. Ultraviolet light from the Sun would not react with CO_2 to form any reasonable quantity of organic matter. So the Earth itself was an unsatisfactory source for life's building blocks. If the planet needed an exogenic source for its supply of organic material from space, comets offered a ready supply.

Two processes, cooling and bombardment, were taking place simultaneously in the Earth's earliest time. The molten planet had to cool sufficiently for a crust to form. During the accident that formed the Moon, Earth lost that crust, but it started to re-form within a few weeks. Was the Earth being pelted by comets after this? If the comet shower had all but stopped, then comets could not be much of a source for life no matter how rich in organics they were. To get an idea of the timing of the period of bombardment, we can explore the craters on the Moon. Unlike the Earth, whose process of weathering and erosion has eliminated the craters from this period, the Moon keeps her record. By studying the crater record there, we know that the Earth-Moon system was being bombarded by cometary impacts up to 3.9 billion years ago, long after the Earth's crust hardened.

If we only consider the comet hits during the period of late heavy bombardment and after, when we are certain that the Earth had a permanent crust, there still would have been enough striking comets to provide for a substantial portion of Earth's water supply and other organic material.

UNITED COMET SERVICE

Invoking comets as the source of organic material begs the problem of how the delivery was accomplished safely. If an express delivery company offered fifteen-minute worldwide service by launching your parcel from the sender and then shooting it through your roof at forty miles per second, you would probably not use their service more than once. A comet hitting the Earth at that velocity gets so hot that virtually all its precious organic material is incinerated.

The key word is "virtually." In a high-speed comet impact, the comet disintegrates in the atmosphere, releasing the organic materials, some of them intact. It is also possible that a comet's organic materials could survive if the comet were moving slowly enough and if it landed at the bottom of an ocean. If the comet collided at a sharp angle, having to pass through a greater amount of atmosphere, and then came to rest in an ocean, some of its organic materials might survive. According to Cornell's Paul Thomas, during the period of greatest bombardment, Earth's atmosphere was ten to twenty times denser than it is now—dense enough to slow a small comet considerably before it hit, possibly enough so that the impact temperature would not be so high as to vaporize its organic materials. The process of using a planetary atmosphere to slow down is called aerobraking, and was successfully used on July 4, 1997, when the *Mars Pathfinder* spacecraft slowed down to land on Mars, and two months later to accomplish the same purpose with *Mars Global Surveyor*. It is also the process that the spacecraft used to slow down at Jupiter in Arthur C. Clarke's science fiction story *2010*.

After a natural experience in aerobraking, an early comet had a very good chance of landing in an ocean, taking less than half a second more to reach the bottom. It is possible that despite the violence of such an impact, the center of the disintegrating comet might not climb above a few hundred degrees Celsius, thus allowing some of the organic molecules to survive.[3]

In a more extreme version of this idea, a comet might strike the Earth at such a slow velocity that some of its organic materials incredibly could survive an impact on land if it hit at a shallow angle, as in a sideswipe. The impact might then form a large crater with a central mound, and it would fill rapidly with melted material from the comet, material consisting of water as well as other organic compounds. Although this would only happen with a very small percentage of impacts, it is a process that could leave organic materials intact.[4]

Did all these impacts really happen? A simple look at the Moon shows that they did. If the Moon was struck so often in its youthful period, it is reasonable to say that the Earth, then only ten thousand miles away, was a target also.

A QUIETER WAY

Suppose your delivery service had an "economy" alternative to shooting your parcel through your roof. Your package would arrive by gently wafting out of the sky, particle by particle. Would that satisfy your delivery needs? Probably not, but Nature does not offer us too many choices. Her slow cometary delivery service works this way: As comets round the Sun, they leave a vast quantity of organic grains in a cloud. This cloud still survives. We see it when the evening or morning sky is lit by a peculiar tepee-shaped glow called the zodiacal light. This beautiful effect, visible with the naked eye on a clear, dark night away from city lights, results from the Sun's light reflecting off interplanetary dust. It is especially strong in the evening sky of late winter.

Although the many particles that inhabit the zodiacal cloud have been there for periods of at least a thousand years, the cloud has had a continuing influx of particles since the beginning of the solar system. It appears that tiny cometary particles from the zodiacal cloud, some of which are from ancient comets, actually arrive on Earth. In 1970, the first balloon experiment to collect dust particles in the stratosphere retrieved volcanic dust as well as some microscopic samples of feathery cometary dust. These cometary "Brownlee particles," named for discoverer Donald Brownlee, eventually drop to the Earth's surface at the rate of a single particle on each square meter of the Earth's surface every day.[5]

In the early eons of the solar system, the zodiacal dust cloud was far more massive than it is now. Over a long period of time, uncountable billions of such grains entered the Earth's atmosphere and floated down ever so gently. These particles added a weighty amount of additional organic materials to the primordial mixture.

HOW DID ORGANIC MATERIAL TURN TO LIFE?

Although there are several characteristics that we ascribe to life, including organization, metabolism, ability to grow and adapt, and even to be irritated, the most decisive is the ability of a molecule to replicate itself. There is a vast difference between a living person and a piece of granite; we do not have trouble distinguishing between life and nonlife at that level. But at the lowest tier, we look for molecules that reproduce through the division of a cell. Life is set up by four bases—adenine, thymine, guanine, and cytosine—that make up the spirals of the deoxyribonucleic acid, or DNA molecule, that is the code of instructions for reproduction of any living thing, be it a bacterium or an astronaut.

DNA has been much in the news lately. Because it is the molecule that singularly defines each of us, it is touted as an ideal personal identifier. DNA is what courts of law use to identify a criminal, a use of DNA that James Watson and Francis Crick hardly had in mind when they began their work on the architecture of the DNA molecule in the fall of 1951. By April 1953, they published their discovery that the molecule was structured as a double helix, a grand staircase of instruction. It is an even simpler molecule, RNA, that was found by T. R. Cech to be the enzyme catalyst for life, a discovery that earned him the Nobel prize in chemistry in 1989. RNA, which stores information, might be the oldest and most basic living molecule.[6]

Comet Hale-Bopp fights for attention with a rocket launch from White Sands Missile Range, and three radio towers. Photo by Tim Hunter.

At its very start, the process that differentiates non-life from life has a number of steps, our understanding of which is still open to hot debate. In one scenario, simple organic materials, like hydrogen cyanide, had to condense to form biomolecules like amino acids. It is possible that this process could have taken place on Earth, in comets, or even in the original solar nebula. We now know that comets contain substances as complex as hydrogen cyanide, and formaldehyde. In 1986, the *Giotto* spacecraft detected a polymerized formaldehyde on Halley's comet.

In a following step, clay minerals catalyzed to form the first nucleic acids, and after a complex process, specific bases were selected and the first RNA formed, helped along perhaps by metal ions. At some point, the RNA began to reproduce itself without the need for the metal ion catalyst. Perhaps the moment when RNA began to reproduce without the need for a catalyst was the point where life began.

Finally, over another long period of time, several new enzymes were selected to form the first DNA molecule. In this story, DNA is the lock on life. Although RNA could reproduce, the more advanced DNA molecule was far more resistant to mutations. With DNA, life on Earth had a good start.[7]

In the fall of 1997, Stanley B. Prusiner received the Nobel Prize for his controversial discovery of self-replicating proteins called prions. He proposed the existence of this protein as a cause of illnesses like mad-cow disease. The idea that a protein can replicate, becoming an infectious agent, is a truly new biological concept. So which is the simplest living molecule? RNA, or a protein like prions? And did RNA, or a protein, become the first living thing on Earth? The debate continues.

COULD THE CHICKEN HAVE COME
BEFORE THE EGG?

If comets contained organic materials, and maybe even prebiotic materials like formaldehyde, is it possible that the process of life actually could have begun on comets? My own guess, shared by most scientists, is a resounding no. But a few scientists do suggest that cometary panspermia, the seeding of life on comets and then transferring it here, is possible. The idea is well known partly because the British astronomer Sir Fred Hoyle, one of the best known scientists of the century, has championed it. The chances for even the simplest of enzymes to form on Earth, he suggests, are so low that these materials could not have formed spontaneously here on Earth. Therefore, the enzymes must have formed elsewhere in the universe and then been transported here directly by comets. However, this part of his argument is self-defeating: chances for life to evolve here are no less likely than for any other spot in the universe, and a very high proportion of the transported enzymes would be destroyed during the impact with Earth. In short, Occam's razor, the philosophical principle that the simplest explanation should be the correct one, applies. Since having comets transport real enzymes would unnecessarily complicate things, that probably isn't how life on Earth began.

If comets did carry life forms here, could cold and flu viruses have come from comets? Although we cannot say distinctly that this never happened, it is extremely unlikely that life was transported in this way. The rhinoviruses that cause colds, for example, are specifically designed to enter the human nose and mouth, and then to thrive once they get inside. Such viruses must have evolved as we, and our noses and mouths, evolved.

In the next chapter, we will explore a variation on this theme—leaving aside the idea that life got its start on comets, but that life might have started on Mars. If that is the case, could comet impacts have transferred simple life forms from Mars to Earth? If this did happen, then Hoyle's idea might, in a very narrow sense, have merit.

An old story takes this idea many steps farther. Why not, suggests a late-eighteenth-century writer, have a civilization on a comet? The idea, going back to 1772, is that the purpose of a comet's tail, which gets bigger as the comet approaches the warmth of the Sun, is to keep the temperature of the comet's surface stable for its inhabitants. As the comet retreats from the Sun, its tail and coma shrink and wrap around it, keeping its inhabitants warmer.[8]

REMINDERS OF OUR HERITAGE

In the rain of comets at Earth's dawn, the seeds for life were planted here. The Earth rarely undergoes such bombardment now, but on November 17, 1999, we might see a miniature version of it. On that night, over someplace on Earth, hundreds of thousands of tiny remnants from Comet Tempel-Tuttle will strike the Earth over a period of a few hours. On this occasion, when the Earth travels through an unusually dense part of a meteor stream, we will experience a spectacular meteor storm. The volley of dust falling from the sky will be a repeat of an event that took place exactly thirty-three years earlier: on the night of November 17, 1966, our Earth plowed through the dusty debris left by the passage of a comet called Tempel-

Tuttle. For a few hours, thousands of people watched the Earth literally plow its way through space as 150,000 meteors per hour fell from the sky.

What causes this trail of debris? It is a single comet, called Tempel-Tuttle, after the two amateur astronomers who discovered it more than a century ago. This comet takes about thirty-three years to orbit the Sun. Year after year, the comet moves along its path. The last great storm was in 1966, and the comet, with its spectacular dust trail, is due back in 1998. Those who are lucky enough to witness the spectacle will get a glimpse of what the Earth was like during its primordial age. As cometary debris rains out of the sky, the remnants of destruction and creation of life will light up the sky just as they did at the dawn of life on Earth.

CHAPTER FOUR

THREE BILLION YEARS AGO

Midnight—with Cole of Spyglass Mountain seated high up on his ladder, his far-seeing blue-gray eye glued to the powerful five-hundred-diameter eye-piece of his telescope. Unnoticeably the refractor followed the planet in its endless flight. The driving clock purred softly, the only sound on Spyglass Mountain—cold and still and fraught with uncanny tensity.

—ARTHUR PRESTON HANKINS, *COLE OF SPYGLASS MOUNTAIN,* 1923[1]

MEET JOSHUA COLE, a fourteen-year-old boy with his head in the clouds and his feet mired in bad luck. Cole is the focus of Arthur Hankins's 1923 novel *Cole of Spyglass Mountain,* a story that lands us in a Charles Dickens–like world in which someone with the fortune of Oliver Twist looks toward the sky and imagines that Mars is a place fit for an intelligent civilization.

Spyglass Mountain opens as Joshua is expelled from school for refusing to let the teacher whip his younger brother. Joshua's father places him in a reform school where the boys are known by numbers, not names: Joshua is 5,635. Joshua's first turn of luck is meeting Beaver Clegg, head of the institution's juvenile department. Clegg is an amateur astronomer who teaches Cole the rudiments of sky watching. Mars is one of the subjects Clegg teaches Cole.

THE RED PLANET

Named for one of the most important Roman gods, the god of agriculture, and later of war, Mars is one of the reddest things we can see in the sky. The fourth planet out from the Sun, Mars is farther from the Sun than Earth. When it is easiest to see, when it rises at sunset and stays in the sky all night, it can be one of the brightest objects in the sky. During this time, it can be as close as 35 million miles from us and twice as bright as Sirius, the brightest star. Mars has a desertlike environment and a thin atmosphere that makes its sky pink. Its temperatures can be surprisingly pleasant; on its equator the

Comet Levy, photographed by the author during a thunderstorm in September 1990. As I set up my Schmidt telescope to take this photograph, clouds thickened and rain began. I shielded the telescope for some fifteen minutes. Then the rain and clouds parted long enough for me to take this two-minute exposure. The clouds, visible at the bottom of the picture, then obscured the comet once again.

thermometer can show a temperature as high as 70 degrees. However, keep the thermometer within a few inches of the surface, for the temperature falls rapidly with each foot of altitude.

Mars has some spectacular mountains, especially Olympus Mons, a volcano the size of Arizona. On Earth, even large volcanoes like the Hawaiian Islands don't get above a certain size because the crustal plates beneath the mountains keep shifting, thus turning off volcanic vents. Since Mars apparently does not experience such shifting, its volcanoes can grow indefinitely.

PERCIVAL LOWELL, MARS, AND JOSHUA COLE

Late in the nineteenth century, the Italian astronomer Giovanni Schiaparelli observed long, straight lines that seemed to crisscross the Martian surface. One of the best observers of his time, he drew these lines as best he could, and then publicized them as *canali*. A few short years after Schiaparelli began his curious drawings of channels, our world was aflame with the possibility that intelligent life existed on Mars. This revelation came not through science but through language. Schiaparelli's *canali* means "channels," but it is spelled like "canals," and that is how it entered the English language.

We've heard of the Lowell Observatory in chapter 2. Its founder, Percival Lowell, was a somewhat eccentric turn-of-the-century amateur astronomer. He came from Boston's famous Lowell family, who donated its name to mystique and poetry such as John Bossidy's 1910 "On the Aristocracy of Harvard":

And this is good old Boston,
The home of the bean and the cod,
Where the Lowells talk to the Cabots
And the Cabots talk only to God.

Schiaparelli's ideas dominated Lowell's mind and soul, and he erected a mighty twenty-four-inch refractor in Flagstaff, Arizona, where he observed the red planet and drew canals that crisscrossed the planet in a complex network. A pacifist who believed in world unity, Lowell used his canal drawings to portray a civilization of intelligent Martians that had joined together to build a system to move scarce water to parched areas of the planet.

Lowell's observations and ideas were scorned during his lifetime, but the exciting possibility of Martian life whetted the popular imagination for many years. We return to our fictitious friend Joshua Cole at the height of the frenzy about life on Mars. As he leaves the reform school to head west, his trip begins well; each evening in a different town, he sets up the telescope and shows the stars to passersby for a fee. Then he hops on a freight train, hoping to make it a little farther west before getting thrown off. But one night, the telescope, his only remaining hold on the stars and an income, is stolen and he continues west alone. Finally, Cole reaches California, where he homesteads on a tall mountain under the clearest, darkest sky Joshua has ever seen. Believing Lowell's theory that a Martian civilization has built a system of canals, he goes a step farther. Obtaining another telescope, he starts a search for a pattern that would be unmistakably artificial.

Back east, Joshua's evil father learns that his son is about to receive a huge inheritance, and arranges for someone to prevent the boy from receiving it, at any cost. All the while, Mars is approaching its closest point to Earth in many years, and on the night of June 18, 1922, Joshua peers through his telescope and tries to see some detail on the planet's surface. (Although *Spyglass Mountain's* author left the year out of his book, June 18, 1922, the year before publication, was indeed the best night to view Mars for many years. It must have been a beautiful sight, with Mars shining near bright red Antares. Planet and star would have looked like a pair of bloodshot eyes.)[2] For anyone looking through a telescope at Mars, seeing detail depends on a steadiness of two atmospheres, one over Mars and another over the observer's telescope. Often, Mars looks like a shimmering piece of Jell-O. But on the night of June 18, the "seeing" improves as the planet climbs the sky. Finally, near midnight, there comes a magic moment when the air is rock-steady. Mars shines forth, its canals bearing the unmistakable pattern Cole is looking for, a pattern that could only have been made by a global civilization. But at that very instant, his father's operatives start shooting at his observatory dome. Before he is able to confirm his sighting, a bullet strikes him. We leave Joshua, for now, lying on the observatory floor.

ACTUAL EVIDENCE FOR LIFE

Talk of canals, Martian unity, and invasions was finally laid to rest during the flyby of the American spacecraft *Mariner 4* during the summer of 1965. MARS AT NOON: NO CANALS screamed one newspaper headline. *Mariner's* pictures showed such a bleak, crater-ridden surface that the story of life there seemed definitively over. During the next decade, several spacecraft continued to chart the red planet. In the fall of 1971, *Mariner 9* arrived in the middle of a planet-wide dust storm and settled into an orbit around Mars. Its cameras worked properly, but they recorded absolutely no surface detail through the blinding dust. As the dust finally subsided, the craft looked down on the red planet and discovered Olympus Mons. It also looked at a canal-like feature called the Coprates canal by visual observers. To *Mariner's* close-up eye, this was no canal but a gigantic canyon as long as the distance across the United States and wider than the length of Arizona's Grand Canyon. Now called Valles Marineris, this enormous feature seems a remnant of an ancient process of water erosion.

On July 20, 1976, two *Viking* spacecraft landed on Mars. These craft were named *Viking* because they were meant to explore a new world and look for signs of life there. They captured haunting scenes of a red planet with a pink sky, and took precise measurements of the content of the Martian atmosphere. Beautiful, but disappointing. One goal was to answer the haunting question: Is there life on Mars? After the two landers collected and analyzed soil samples for months, scientists studying the data concluded that the surface within reach of the landers' arms was absolutely sterile.

There the story rested. For the next two decades, no spaceship made it to Mars. In 1993, the American spacecraft *Mars Observer* was making its approach to the red planet when Earth-based flight controllers turned on its braking rockets. The craft fell silent and was never heard from again. Although NASA never definitely figured out the cause of the failure, the agency suspects that a fuel explosion destroyed the craft. For two years, Mars was virtually ignored in the news media. Then, in the summer of 1996, the long story of Earth's fascination with Mars took a very new turn.

As millions watched on worldwide television, President Clinton announced that evidence had been found that 3.6 billion years ago, not too long after the end of the late heavy bombardment, Mars might have had simple forms of life.

MARS ROCK ALH 84001

On December 27, 1984, a team of meteorite hunters in Antarctica came across a greenish rock the size of a potato. The rock was immediately thought to be a meteorite, since stray rocks of other origins are not likely in the ice. As meteorites fall in Antarctica, the ice slowly carries them toward a region called the Allan Hills, where they can remain buried for thousands of years. As the ice melts, the meteorites are finally exposed to sunlight and possible retrieval by alert hunters. The team on which Roberta Score worked has found many, but this one, she says, was the greenest meteorite she had ever seen. The team labeled this specimen ALH 84001, the first find of the new year. Meteorite geologists knew the rock was special, but not at first how special. A full decade passed before the rock was identified as having made the trip from Mars. Scientists were certain "beyond a reasonable doubt" that ALH 84001 was from Mars, because it contains tiny pockets of atmosphere virtually identical to the atmosphere that the *Viking* spacecraft studied on Mars in 1976. These pockets, created when the rock originally formed on Mars, are "airtight" enough to lock in small amounts of Martian air virtually indefinitely. When the meteorite finally was brought to a lab, scientists could then penetrate the pockets and measure the composition of the air inside.

The extreme age of the meteorite—4.5 billion years—makes it by far the oldest rock ever recovered from Mars. (The other eleven known Mars rocks are not more than 1.3 billion years old.) This rock almost certainly hardened from magma as part of the original crust that formed when Mars cooled. The rock is so old that it can tell us directly about the period of late heavy bombardment that sent objects slamming into all the inner planets of the solar system. While still on Mars, the rock led an eventful life. One impact fortuitously happened not far from the rock, cracking it. The crash took place between 4.0 billion and 3.8 billion years ago, during the bombardment period. Later, the rock spent at least one episode under water rich with carbon dioxide. As the water seeped through, tiny pellets of carbonate formed along the rock's fissures. There the rock stayed as Mars dried up completely. Then, 15 million years ago, a second impact thrust the rock into space, where it orbited the Sun in ever-changing paths. It probably experienced several near misses with Earth. Finally, thirteen thousand years ago, the rock fell onto the Antarctic ice. Buried there all this time, it was slowly pushed toward the Allan Hills, where the ice melted enough to reveal it.

One surprise is that water formed on Mars in the first place. We do know that water was abundant on Earth at the time, almost 3.5 billion years ago, and the first sausage link–shaped microfossils were appearing in Greenland and elsewhere on Earth. But on Mars, from the time of the first spacecraft visits, absolutely no evidence of water has been found. Now Mars sends us a rock that might be telling us that in the distant past, water may have flowed freely on the red planet.

Other scientists dispute the NASA team's date of 3.6 billion years that coincides with a period when water was probably flowing on Mars. Their dating makes the carbonates much younger, between 1.4 and 1.3 billion years ago, long after water was supposed to have vanished from our neighbor world. If the water inflow idea "holds water," either it happened some 3.6 billion years ago, or water was still flowing on Mars 1.4 billion years ago. Further study of the specimen will undoubtedly resolve this dispute.

When *Mars Pathfinder* landed on Mars on July 4, 1997, it began photographing a surreal panorama. Huge boulders strewn about told a story of a long-gone major flood, during which water gushed through the area with enough force to push the heavy boulders around. A strong force, indeed. Did *Pathfinder* land in an ancient riverbed, or was the flood a major one after some sort of storm? Whatever the cause of the flood or its duration, we now have considerable evidence that Mars, at a time far back in its past, must have had liquid water—and lots of it.

The evidence from both *Pathfinder* and the rock called ALH 84001 makes a strong case for water, and where there was water, there could have been life. But an accurate date for the penetration of the carbonate-rich water was crucial to the success of the NASA team's interpretation. The water that flowed through the rock was rich in carbon, and left tiny globules of carbonate with Oreo-cookie-like bright- and dark-layered rims. Made of calcium and magnesium carbonate, these globules resemble antacid pills. It is in and near the carbonate globules, each no larger than $\frac{1}{100}$ of an inch, the size of a punctuation mark, that evidence for life has been found by a NASA team led by David S. McKay and Everett K. Gibson. Using a scanning electron microscope, the team found strange groups of long, hot dog–shaped forms in the carbonate globules or near them. Could these be microfossils? Although they do resemble ancient Earth microfossils, they are about one hundred times smaller.

Even though the appearance of the fossil-like structures raises hope that the science team found life, they are probably the evidence chain's weakest link. The second line of evidence involves tiny grains of magnetite and iron sulfide. It is unusual to see these compounds existing together, and very unusual to see them together with carbonate globules. These grains are evidence for life because they are typical of the decay products of some Earthly bacteria. Not everyone agrees: according to a research group led by geochemist John Bradley, a volcanic eruption, and not any living thing, could have produced the magnetite.

Another line of evidence is the PAHs. An acronym for polycyclic aromatic hydrocarbons, PAHs are organic molecules. PAHs are also by-products of decaying organic matter. Their presence in the rock from Mars is not necessarily strong evidence for life, since these molecules might have existed in the ancient nebula that condensed to form the solar system, and they have been found in other meteorites. So why should these particular PAHs be so indicative of life? Or the PAHs could be simple contamination from the Antarctic ice the rock was embedded in for thirteen thousand years, as geochemist Jeffrey Bada suggests.

If the biological answer is correct, the temperatures in which the carbonates formed could not have been higher than 300 degrees Fahrenheit. If the temperature was much hotter, then a nonbiological explanation must be found, since life cannot start or exist at that temperature. Some scientists suggest that the temperature might have been as high as 1,100 degrees. But in March 1997, several months after the original announcement, scientists at the University of Wisconsin presented evidence that the carbonate grains did not form rapidly at high temperatures, but instead molded slowly at temperatures low enough for a biological process to occur.

Each of the three lines of evidence—shapes that look like microfossils, iron-rich grains, and PAHs—is not sufficient to conclude that the Mars rock contains evidence of life. But David McKay senses that all three taken together converge on a biological interpretation.

A SECOND LIVE MARS METEORITE: EETA 79001

Just when we thought we had heard it all, at the end of October 1996, a team led by Ian Wright of England's Open University found that a second Martian meteorite was laced with an organic component called carbon 12. This enriched carbon is most easily explained as the remains of activity from microbes. This rock, called EETA 79001, is far younger than ALH 84001. Made of basalt, the meteorite is a fragment of volcanic lava that flowed on Mars some 180 million years ago, becoming a part of the Martian landscape. Just 500,000 years ago an impact thrust the rock into space, where it began its own orbit around the Sun. Slowly, it moved closer to Earth, and it eventually fell in Antarctica.

WHY DOES MARS LOSE SO MANY OF ITS ROCKS?

Whenever a comet or asteroid smacks into a planet, it digs itself a large hole in the ground at least twenty times the impactor's own diameter. Because the resulting crater is formed extremely fast, in a few minutes enormous amounts of debris are tossed upward. Above the Earth, virtually all of these rocky remnants soar through the atmosphere and come crashing back to the surface again. On Mars, although the impact force is the same, the planet's gravity is much less. Instead of returning to Mars, the fragments achieve escape velocity and hurl themselves into space. Once that happens, it is reasonable to expect that many fragments—not just a few— will find their way to Earth, the next stop inward toward the Sun. The fragments reach orbits that take them out by Mars, then a short way toward the Sun. Each successive orbit brings the fragments a bit closer to Earth. Some will crash to Mars again, others to Earth or the Moon, and still more to Venus or even Mercury. There are lots of Martian meteorites here, waiting for some alert searcher to find them. Most will be found in Antarctica, where they stand out clearly against the permanent white. But who knows? Maybe a schoolhouse wall contains a Martian rock. The American poet Robert Frost thought of something like that:

> Never tell me that not one star of all
> That slip from heaven at night and softly fall
> Has been picked up with stones to build a wall.
>
> Some laborer found one faded and stone cold. . . .[3]

A PROCESS OF SCIENCE

A few months after the life-is-possible announcement stunned the world, I was invited to participate in a colloquium at Montreal's McGill University, sponsored by

NASA team member Hojatollah Vali. In front of a packed auditorium, the panelists answered question after question. I think we were all amazed at the interdisciplinary nature of this discovery. There was geology, a lot of biology, considerable chemistry, and physics. But near the end came a question about politics. "You seem to be standing on a precipice," the student began. "Why didn't you wait till more solid evidence comes in? Now, if new evidence proves you're wrong, your whole team will have egg on your faces."

I decided to tackle that one. "No," I said, "this team will never fry an egg. You are witnessing today precisely what science is all about. The amount of evidence they have is enough for the team to propose an interpretation to the scientific community, which they have done, and then throw the question, and the samples, open for others to work on. If some other scientist comes up with a better explanation of these events, that's fine. There is no embarrassment, for that is how science progresses. That is how we learn."[4]

"Rock 84001 speaks to us," said President Clinton, "across all those billions of years and millions of miles. It speaks of the possibility of life. If this discovery is confirmed, it will surely be one of the most stunning insights into our universe that science has ever uncovered."[5] The NASA team's research will eventually lead, I believe, to the direction that an infant Mars did support simple life forms, and even that it still may support life at this level. But whether or not ALH 84001 is a Rosetta stone for life, it has brought the age-old question of life on Mars to a new and thrilling level.

FAMOUS IN A NIGHT

Let's return a second time to Joshua Cole, since we left him hovering near death on the observatory floor. Depressed by a suspicion that no one can confirm his discovery, Joshua knows that Mars will not be so close to Earth again for many years. But then the reports come in: an observatory "has photographed figure unknown American observer saw on June Eighteenth." Joshua seems reborn as he gazes at the newspaper headlines: "SCIENTIFIC WORLD IS ASKING BREATHLESSLY: IS MARS A LIVING PLANET?"

What Joshua Cole does not know is that sometimes life imitates art. Three-quarters of a century after *Spyglass Mountain* first appeared, I looked up asteroid number 5,635. A small rock in orbit around the Sun between Mars and Jupiter, it had never been named. I wrote Bobby Bus, its discoverer, and suggested the name Cole, for the number he had in the boys' reform school. Bobby enthusiastically agreed and proposed the name to the International Astronomical Union's Minor Planet Names Committee. In December 1997, Joshua Cole got a world named for him. The newly named asteroid would have added to the joy Cole felt as he sat up in bed and continued to read the headlines: FIRST TO REPORT DISCOVERY, a headline called out, COLE OF SPYGLASS MOUNTAIN FAMOUS IN A NIGHT.

CHAPTER FIVE

SIXTY-FIVE MILLION YEARS AGO

Contemplate all this work of Time,
  The giant laboring in his youth;
  Nor dream of human love and truth,
As dying Nature's earth and lime . . .

—TENNYSON, *IN MEMORIAM*, 1850[1]

THE EARTH IS A book, its story written in page after page of rock. Its language, the tongue of geology, is not difficult to understand for people trained to see what layers of rock tell them. For 65 million years, a crucial page in the story of the Earth lay waiting for someone to read it. It was waiting in a worldwide layer of rock as thin as a credit card, as well as in a gigantic crater buried under Mexico's Yucatán Peninsula. After all those millions of years, in the late 1970s, Walter Alvarez and his father, Luis, opened the book, read the page, and in a beautiful mountain region near the medieval town of Gubbio, Italy, found buried treasure.[2]

The riches lie in a pinkish rock called *Scaglia rossa* limestone. Laid down in the deep water that covered this part of Italy 65 million years ago, the rock was undisturbed by anything but the gentle flow of ocean currents. And nowhere on Earth is it displayed better than in the hills near Gubbio. This layer became interesting to Walter Alvarez as he studied how it recorded the ancient direction of the Earth's magnetic field. The Earth is a magnet with two poles, north and south, that point not far from the geographical north and south poles. If the Earth did not have a magnetic field, compasses would never have been invented, for they would have been useless. In 1997, the *Mars Pathfinder* spacecraft confirmed that Mars has such a weak magnetic field that compasses do not work there. The Earth has always had a magnetic field, although on occasion its direction has shifted. As continents moved and rotated through billions of years, a compass planted on a rock would point to the direction of the Earth's magnetic pole at the time the rock was formed.

Geologists have a way of measuring the rock to determine where a compass would have pointed had one been there. They call this procedure measuring the rock's "fossil compass." In this way, the scientists determined that the pole pointed in one direction when most of the samples were formed, but in the opposite direction for others. Clear evidence for reversals of the Earth's magnetic poles was what heightened Alvarez's interest in the rocks near Gubbio.

For several summers, Alvarez worked on these beds, collecting sample after sample to improve his understanding of the magnetic field changes. Since he was concentrating on the Cretaceous limestone, he learned to identify the sharp boundary, within the limestone, between two major geological periods of time, the Cretaceous and the Tertiary.

Geologists measure time on Earth in millions of years, but also in specific periods that are characterized by the emergence or disappearance of specific life forms. The history is broadly divided into major eras: the Precambrian, whose Archean period stretched from the formation of the first solid rocks on Earth to the appearance of the earliest life forms, and whose Proterozoic period began when fossils first appeared in large numbers some 2.5 billion years ago. The Paleozoic (ancient life) era began 540 million years ago and is divided into several periods. The Mesozoic (middle life) era began and ended with the dinosaurs. It is divided into three periods, the last of which is called the Cretaceous, named after the famous white chalk cliffs of Dover, England. The Cretaceous borders on a new era, the Cenozoic (recent life), of which the Tertiary is the earliest period.

THE GEOLOGICAL TIME SCALE

Relative Duration of Geological Intervals

Era	Period	Time Began, Years Ago
Cenozoic	Quaternary	2 million
	Tertiary	65 million
Mesozoic	Cretaceous	136 million
	Jurassic	205 million
	Triassic	225 million
Paleozoic	Permian	280 million
	Carboniferous	315 million
	Devonian	345 million
	Silurian	395 million
	Ordovician	430 million
	Cambrian	540 million
Precambrian	Proterozoic	2.5 billion
	Archean	4.6 billion

I learned about eras and periods at Acadia University, in Nova Scotia, where I was lucky enough to have George Stevens as my professor for Geology 100 in the fall of 1968. He introduced us to the two major geological schools of thought, catastrophism and uniformitarianism. Made popular by late-eighteenth-century French scientist Georges Cuvier, the catastrophist school held that changes on Earth happened suddenly and violently. Considering the violence of the French Revolution, which was going on during Cuvier's lifetime, it is not surprising he attributed the habits of humans to those of the Earth. The more accepted uniformitarian school, on the other hand, saw the Earth's story as a long, gradual march forward. But George Stevens told me also of a man who was changing the focus of geology in the direction of the other school. Perhaps the Earth was still uniformitarian, but was the long, steady march punctuated by some abrupt changes?

Continuing their study of the Gubbio rocks, the Alvarez team wondered about the subtle border between major eras of the Mesozoic and the Cenozoic on Earth. As the *Scaglia rossa* limestone was being laid down peacefully underwater, something catastrophic must have been happening above water. The same kind of rock was being deposited, but suddenly vastly different life forms were preserved in it.

On first glance, the thin clay layer separating two limestone beds did not seem unusual. It is called the boundary layer, but no bells and signs herald its presence in the limestone. Although the boundary layer was a little thicker, several other clay layers separated the limestone beds. The profound difference was in the life forms preserved in stone below and above the boundary. In the Cretaceous beds were fossil shells of foraminifera, one-celled marine animals that lived out their lives on the sea floor. The foraminifera were as large as grains of sand and very numerous in the uppermost limestone beds of the Cretaceous. In the younger limestone, there were far fewer foraminifera fossils, and those that did exist were much smaller.

These tiny creatures became virtually extinct at the same time as the huge dinosaurs. It was time, the Alvarez team concluded, to study the boundary clay itself, to see if anything preserved in it might answer the riddle of what appeared to be a sudden, mass extinction. Back in Geology 101, we were taught that even though extinctions looked sudden in rock layers, the decline of a species population might still spread out over millions of years. But thanks to his earlier work on dating the Earth's magnetic field reversals, Alvarez concluded that whatever happened at the boundary resulted in the mass extinction of more than three-fourths of all the species of life in a period no longer than 100,000 years. Alvarez was about to learn that the period of extinction might have been as short as a few months.

By 1976, Alvarez's work focused on the thin boundary layer. Joined by his father, Luis, an experimental physicist whose work on subatomic particles had won him the Nobel Prize eight years earlier, Walter searched for a clue to the mystery. The answer might be, Luis proposed, in a search for one of the rare platinum-group elements, especially iridium. A nuclear chemist at Lawrence Berkeley Labs, Frank Asaro, began a months-long evaluation of the clay for evidence of these elements. In June 1977, he hit the jackpot by detecting a much greater concentration of iridium than anyone had hoped for. By a high level of iridium, Asaro didn't mean that the precious metal was shooting out of the boundary layer in amounts that would enrich the world. The amount he detected was nine parts per *billion,* actually an extremely tiny amount—the level was just many times higher than expected. That much iridium might have come from a long period of sustained volcanic action. However, the evidence indicated that all the iridium fell in a very short period of time, something that would occur only during a brief cataclysmic event.

Could the dinosaurs have become extinct before the boundary was reached? Fossil collectors in the past decade found large dinosaur fossils almost to the boundary layer, and not one above it. Although it is possible that no dinosaur witnessed the last day of the Mesozoic era, it seems unlikely that the demise of a group of creatures that ruled the world, in one form or another, for 150 million years, was not related to the cataclysm that took so many other lives.

The study of smaller fossils paints a clearer picture.

The foraminifera at Gubbio were plentiful—thriving by the thousands per cubic foot of rock—right up to the very boundary layer, and then virtually stopped. Ammonites, creatures that resemble the present-day chambered nautilus, were also plentiful right up until the boundary, and then vanished. At this point in our discussion, we meet again that logical rule of scientific thought called Occam's razor. We see that smaller creatures clearly were present right up to the boundary, and the larger ones, of which there are far fewer representatives, left fossil remains or footprints fairly close to the boundary. Demanding the simplest explanation, Occam's razor says that the same event that wiped out the smaller, more durable creatures probably destroyed the larger ones, too.

WHERE'S THE BEEF?

Intriguing as it was, the discovery of nine parts per billion of iridium is not likely to make newspaper headlines or talk shows. If an object struck the Earth at a velocity of thirty or forty miles per second, it must have left an impact scar in the form of a crater—and a big crater to boot. This event would have explained the layer of iridium. The big mystery during the decade following the iridium announcement centered on where, or even if, there was such a crater.

The lack of obvious hundred-mile-wide craters on land prompted a focusing on an ocean-floor impact, and a search began for the crater there. A major hurdle to overcome was that the ocean floor is constantly being replaced by a process called subduction, in which the floor repaves itself, destroying evidence of older features. About a fifth of the floor has been subducted in the last 65 million years. Thinking that the fate of the crater had been sealed that way, some geologists despaired of ever finding it.

By 1984, Bruce Bohor, a U.S. Geological Survey geologist, found evidence that boundary sites throughout the Rocky Mountains showed evidence of shocked quartz. Quartz is one of the most durable minerals known. Nevertheless, this quartz showed shock waves, as if it had been deformed like a wave of water. Even the strongest earthquakes cannot deform quartz in this way. The shocked quartz significantly narrowed the search for the crater's origin. Quartz is commonly formed on land but not in deep water; thus it is virtually absent on the ocean floor. If so much quartz was affected, the impact site must have been on, or near, a continent!

The story's focus next moves to Haiti, where for many years geologist Florentin Maurrasse knew of a site near Beloc, Haiti, where rocks at the Cretaceous-Tertiary boundary are exposed. But this place, it turned out, was unlike the other boundary sites. Most of the other sites contained a thin layer of clay to mark the boundary, but this one contained sand. In 1985, geologist Jan Smit suggested that the bed may be the remains of an impact-generated tsunami, or tidal wave. These giant ocean waves, which have nothing to do with tides, are best known to follow earthquakes. An impact from space, however, would have generated tsunamis so high that areas hundreds of miles inland might have been flooded. Could the sandy debris have been carried aloft and redeposited in seconds by a mile-high tsunami? In 1988, the Canadian geologist Alan Hildebrand, of the University of Arizona's Lunar and Planetary Lab, looked closely at another known boundary site, this one along the Brazos River in Texas. If the boundary layer there was filled with deposits from a tidal wave, then the wave must have come up from the south through the Gulf of Mexico, which then as now was partly landlocked. A tidal wave through the Gulf must have had its beginning at a site not too far away. In addition to the iridium and shocked quartz, Hildebrand saw something new at this Texas site: tektites, forms of glass made from rocks that melted rapidly and then cooled quickly as they returned to the ground. Tektites are the result of either an intense volcanic explosion or an impact. In Haiti and Texas, the boundary layers seemed so rich with material like this that Hildebrand concluded that ground zero was close by. Of the possible sites Hildebrand considered, the most tempting was a buried area centered at Chicxulub, in Mexico's Yucatán Peninsula.

AN OIL COMPANY IN MEXICO

In one sense, the search for the crater was actually over before it even began. Just one year after the iridium announcement, Glenn Penfield and Antonio Camargo, geologists working for Mexico's Petróleos Mexicanos (abbreviated as PEMEX) oil company, suspected that a hundred-mile-wide feature previously thought to be a buried volcano might actually be the buried remains of an impact crater. As is typical for oil company work, Penfield's results went unpublished for some time, and were essentially unnoticed when he did present them at a petroleum geology meeting that was not attended by specialists in impacts.[3]

It is intriguing that Penfield's discovery did appear as a news note in *Sky and Telescope* magazine as early as March 1982, complete with his suspicion that it might be related to the boundary layer and the great extinction at the end of the Mesozoic.[4] For the next eight years, 100,000 copies of the magazine's news note sat unnoticed in libraries.

By the spring of 1991, Penfield's work was noticed, and rapidly became part of the impact geological landscape. A large number of cenotes, deep ponds filled with fresh, clear water, form a semicircle that matches the edge of the great crater buried far below. Perhaps, as the rim of the crater collapsed, the limestone above it weakened in small areas, forming the sinkholes.[5] Satellite images also showed some evidence of a circular feature. Finally, a series of rocky cores drilled out of the deep crater by PEMEX geologists provided convincing evidence that the smoking gun, actually the smoking bazooka, had been found.

THE EJECTA BLANKET

The boundary detective story is so complex, and involves so many scientists and their colleagues, that it is impossible to credit more than a handful in these pages. Most important, the work is nowhere near complete. The great difficulty of drilling three thousand feet underground to find the crater itself means that scientists need to work with Nature to uncover her evidence. Indeed, Nature provided the needed materials in the form of an ejecta blanket. Filled with enormous amounts of material that was crushed, shocked, and otherwise mutilated, the blanket was thrust out of the impact area and tossed elsewhere. Nowhere is the landing area of this blanket more accessible than in the small country of Belize. Tucked into the southeast side of the Yucatán Peninsula, this ancient land was once a part of the great Mayan civilization. That was more than a thousand years ago. Sixty-five million years ago, the land of Belize was underwater. Less than twenty minutes after collision, the area was buried under a hail of huge boulders and debris. For here in Belize, the boundary layer is not an innocuous strip of clay enriched with iridium; here it is a hundred feet thick, filled with many types of boulders. Heaved out of the impact site, these rocks are a message direct from ground zero.[6]

WHAT HIT US?

In a sense, it does not matter whether the impacting body was a comet or an asteroid. Both objects could deliver the same amount of iridium, and inflict the same kind of damage. The size of the crater does not tell us much, since we do not know how fast the object was traveling when it hit. The main difference between a comet and an asteroid, at least regarding their ability to inflict damage on Earth, is the velocity. An asteroid on a collision course with Earth will probably be moving in the same direction as Earth and come from a part of the solar system not farther than the asteroid belt between Mars and Jupiter. On the other hand, a comet approaches us from a great distance, from at least as far out as Jupiter's orbit to as far as the Oort cloud. That means that a comet's impact velocity would have to be much higher than that of an asteroid. Moreover, a comet can be coming at us from virtually any direction, meaning that it might impact the Earth head-on, increasing the impact velocity to as high as forty or fifty miles per second, fast enough to cross the state of New York in under ten seconds.

The two strongest arguments in favor of a comet are size and population statistics. Of all the asteroids now following orbits that intersect the orbit of the Earth, none are big enough to create the impact damage discovered under the Yucatán Peninsula. The largest Earth-threatening asteroid is 1,627, called Ivar, and it is five miles across. Should it hit us, the resulting crater would nowhere near approach the size of Chicxulub. Chances are that 65 million years ago, while the individual asteroids that were in Earth-crossing orbits were different, their size distribution was about the same as it is now.

The distribution of comets is different. We know of many comets that are miles wide that have neared the Earth in this century alone, including Halley's comet, Comet IRAS-Araki-Alcock in 1983, and Comet Hyakutake in 1996. These comets and others are large enough to cause the kind of damage that was seen at the boundary. Armed with all this evidence, let's close with a look at a night when hell touched Earth.

DINO'S LAST STAND

A peaceful, muggy day draws the Mesozoic to a close. A *Tyrannosaurus rex* roams the ground in search of easy prey, its huge weight shaking the ground like a series of small earthquakes. As the Sun sets in an ocean bath to the west, a lone pterodactyl flies by in the distance, its silhouetted form an invitation to peaceful twilight.

But night comes without darkness. A comet is shining so brightly that the big dinosaurs cast shadows from its light as they move. The comet moves toward the southern horizon and sets. With the comet below the horizon, the sky finally darkens to a view of hundreds of thousands of meteors falling. For a moment, the scene is silent and surreal. Then the Mesozoic ends as the comet breaks through the upper atmosphere and destroys the planet's ozone layer. With a deafening sonic boom, the comet races through the atmosphere as if it weren't even there. Seconds later, with the force of 100 million hydrogen bombs, it slams into Earth just off the coast of present-day Yucatán. Virtually every rock within five miles of ground zero is instantly vaporized. The Earth trembles as a force-12 quake topples any dinosaur trying to stand up. In less than a minute, a mighty shock wave gouges out a crater a hundred miles wide and twenty-five miles deep. Another shock wave tears the rest of the comet apart, and as its remnants plow into the Earth, they vaporize. As the crater's outer walls begin to collapse, rocky material builds a high central peak that soon collapses into a set of concentric rings.

Mile-high tsunamis rush upward from the point of impact, tearing across the Gulf of Mexico and flooding present-day Florida and the southern tier of states. A fireball of hot gas, visible for thousands of miles, rises high into the atmosphere. Millions of tons of rocky debris and dust billow upward in a gigantic cloud. All over the world, the sky is filled with a storm of meteorites as the rocky material dredged up from the Earth strikes the ground again with enough violence to tear it up. Any dinosaur in view of the sky would feel temperatures as high as an oven set to broiling. Ground fires ignite and quickly spread around the world, burning on for months. Fine dust settles high in Earth's atmosphere, as it does after a volcanic eruption, only much, much thicker. The sky is absolutely black, and for over six months, there is no sunlight whatsoever, anywhere on Earth. What rain falls is charged with large amounts of sulfuric acid. It is as dark as a photographic darkroom.

For hundreds of miles from ground zero, virtually all life is eradicated. Within a few hours, all of present-day Mexico and most of the United States are rubble. Unable to survive this terrible event, the large dinosaurs probably don't last more than a few days or weeks. Within a year or two, many plant and animal species are gone. As the sky finally clears after more than two years of "impact winter," temperatures begin a slow rise as a centuries-long global warming sets in. In a few seconds of hell, one of the most successful periods of life Earth ever knew ends forever.

CHAPTER SIX

COMETS ARE DOUBLE-EDGED SWORDS

Turning and turning in the widening gyre
The falcon cannot hear the falconer;
Things fall apart; the center cannot hold;
Mere anarchy is loosed upon the world,
The blood-dimmed tide is loosed, and everywhere
The ceremony of innocence is drowned;
The best lack all conviction, while the worst
Are full of passionate intensity.

—WILLIAM BUTLER YEATS, "THE SECOND COMING," 1920[1]

EIGHT MONTHS AFTER THE apocalypse, a terrified mammal peeks out of the small, dusty hole of its underground home. It sees a land transformed. Where once there was a forest, now all is charred wood and utter devastation. The sky is brown and murky, but for the first time since the impact, a dim Sun is seen rising through the mist. The animal is indeed holding infinity in the palm of its hand: in that hazy dawn, it sees the half-billion-year-old course of Paleozoic life on Earth at an end, and a new world take over.

THE DAWN OF THE CENOZOIC

Every time a giant comet strikes the Earth, the dice of evolution are thrown again. By upsetting the apple cart, life gets a chance to diversify. The dinosaurs, large and lanky though they were, had a surprisingly stable reign on Earth that lasted, from one species to another, for 130 million years. That's an eternity compared with the brief 4 million years humans have been here. Had the impact not occurred, the dinosaurs might still be here today, although changed in some minor ways over the 65 million years that have passed until the present. Any child who has seen *Jurassic Park* knows that humans and dinosaurs probably could not coexist: the death of the dinosaurs was a boon for us. Major impacts are good for life, so long as they do not occur too often.

As the Paleozoic edged into the Mesozoic, the first mammals appeared. They probably evolved from strange reptiles with teeth differentiated in relation to their eating habits. These "cynodont therapsids" bridged the reptile-mammal gap. A simple, mouselike creature, the earliest mammal existed at the same time as the earliest dinosaur. A hundred million years later, the great reptiles ruled the world. The mammals were still tiny, and still scampering about. Apparently, the mammalian role in life's web was to supply themselves as food for the big dinosaurs. The mammals of the Mesozoic subsisted on insects, worms, and organic remains in soil and in rotting logs.

Then, without warning, the comet came. In a few months, most of the dinosaurs vanished, and probably every last one of them was gone within a few years. Gone were the three orders of class Reptilia that the dinosaurs filled: Saurischia (including *T. rex*), Ornithischia (with *Triceratops*), and the flying Pterosauria. A huge number of less famous life forms also didn't survive. A recent study led by Peter Sheehan of the Milwaukee Public Museum divided land life into two subsets, the land species that fed on live tissue, and the freshwater species, which had the option of feeding on detritus. The study found that 88 percent of the species of land creatures became extinct, but a similar percentage of the freshwater creatures, which fed as the mammals did, managed to survive. Since dinosaurs depended on the living tissues of plants or animals, they did not survive. For marine life, the statistics are similar. Species depending on a food chain involving phytoplankton (plankton carrying out photosynthesis) were wiped out in far greater numbers than were species that could feed on washed-away organic material.[2]

Slowly at first, then much more aggressively, the mammals grew to fill the niches left vacant by the departed dinosaurs. Had the comet not come, it is safe to say, our entire class Mammalia would still be no larger than rats. The same comet that probably wiped out the dinosaurs 65 million years ago is the comet we have to thank for making our world the haven for mammals we enjoy today. Our lives would not exist had it not been for the coincidence in space and time that brought a planet and a comet together.

A NEW BIOSPHERE

If our frightened mammal could have lived for a few million years after the impact, it would have witnessed a vast change in the habits of its descendants. The mammalian body size increased as its class of animals imitated the eating habits of the vanished dinosaurs. For the fifth time in the 500-million-year history of complex life forms on Earth, life on Earth changed direction in a big way. Prior to each of the mass extinctions, the biosphere enjoyed a long period of peace and stability, with only gradual change. These long periods are called ecologic evolutionary units and are characterized by a gradual evolutionary process in both plants and animals.[3] Each mass extinction was followed by a burst of speciation and rapid adaptation lasting a few million years, and then followed by a new, long-lasting evolutionary unit.

We are now in the midst of an ecologic evolutionary unit. Such units are easy to picture. A museum display, or a painting, of a landscape several million years old contains creatures that are not that different from the animals we are familiar with today. Evolution has taken place slowly and peacefully through that time. However, a scene containing the late Cretaceous dinosaurs like *T. rex* doesn't look at all like any modern view. These animals are similar to the creatures of the earlier unit. With the comet impact 65 million years ago, the whole focus of life changed.

That focus will change again. Sometime in the future another comet or asteroid, several miles across, will collide with Earth. With present technologies, there is nothing we can do to prevent such a catastrophe; changing the orbit of a ten-mile-wide comet hurtling toward us at a high velocity is, one scientist insists, like trying to move a tank with a popgun.

WOULD WE SURVIVE A MASS EXTINCTION?

According to scientists Peter Sheehan and Dale Russell, humans should be able to survive the next mass extinction, just as their mammalian ancestors did 65 million years ago.[4] After the global fires have ended and the darkness lifted, after the tsunamis have subsided and the acid rain dried up, some humans would probably still be around to put up with a planetary mess. As a species, humans might still exist, but well over half their population could well perish.

Most large animals and livestock would be gone. Thus, the surviving humans would have to become vegetarian. Smaller animals like raccoons, who depend on detritus-based food chains, might survive, as would some birds, lizards, and insects. Assuming that humanity had enough time to prepare for this catastrophe, there would be worldwide food-storage bins. They would have to be carefully protected both from unauthorized entry and from rodents and insects whose survival would depend on the same stores. Although small independent farms would continue to try to grow food, the utter collapse of civilization would make distribution of foods very difficult. Survival after the holocaust would be no picnic, but it should be possible.

THE COMET'S ROLE IN THE BIG PICTURE

We often think of the Earth's distant past as a single time lumped all together as one long instant. Time, it has been said, is Nature's way of making sure everything doesn't happen at once.[5] Actually, from the time of the primordial cloud, to the late heavy bombardment, to the origin of life, all the way along the panorama of species to the end of the dinosaurs, represents about 99 percent of the Earth's history! By the time of the dinosaurs, the great period of late heavy bombardment was but a distant planetary memory that had stopped relatively quickly 3.9 billion years before. By that time, Jupiter, the solar system's vacuum cleaner, had done its work, its massive gravity clearing the solar system of most of the cometary bullets.

The end of the world of dinosaurs is by far the most famous of the mass extinctions. But hidden in the fossil record is evidence of others:

- The boundary between the rocks of the Precambrian and the Cambrian layers is more than half a billion years old and difficult to study. It is enough, however, to tell us that the Cambrian began with a sudden proliferation of life. After a billion years of one-celled life forms, life suddenly started to get fancy. In the early Cambrian lie the earliest records of creatures with mineralized skeletons, including shelled creatures called trilobites. Also present during this period were chordates, the earliest members of the phylum that eventually included humans. These early chordates, however, did not have hard skeletons and would not have been preserved as fossils. Did an impact, or a series of impacts, launch all these new life forms? Although the evidence is far from conclusive, there is a mild iridium anomaly at the boundary, and there is a fifty-five-mile-wide impact structure at Lake Acraman, in South Australia, that could be dated at the boundary.

- The end of the Ordovician period, some 450 million years ago, was punctuated by the oldest known mass extinction. Two smallish craters, in Brent, Ontario, and near Calvin, Michigan, could be all that is left of a shower of asteroids or comets from that time, although other than timing, there is no real evidence to connect these events. Moreover, a long period of glaciation, which would not be expected after an impact, accompanied this loss of life.

- Near the end of the Devonian period, some 365 million years ago, the second known mass extinction resulted in a major loss of species, including reef-building corals. Some evidence that this extinction is impact-related is in the discovery of impact-formed glass spherules in China and Belgium.

- Almost 250 million years ago, the Paleozoic era ended with the third, and by far the most violent, mass extinction in the recorded history of the planet. A whopping 95 percent of all the species of life were extinguished at that time.

What would trigger such an enormous loss of life? Many geologists think that the Earth itself was a very busy place at the time, its tectonic plates having just brought the continents of Gondwanaland and Laurasia together to form a supercontinent we call Pangaea. As sea level declined, the ocean basins were deepening. Could worldwide massive volcanic eruptions associated with this activity have led to the mass extinction? However, new evidence shows that this greatest of all mass extinctions took place over a period no longer than fifty thousand years.[6] A prolonged period of volcanism should have allowed the extinction to be spread over millions of years. On the Falkland Plateau at the bottom of the South Atlantic Ocean sit two large structures, one of which is two hundred miles wide. Could these be the remains of craters that formed from impacts at the end of the Permian? Or was this greatest extinction of all a result of something other than a cosmic impact?

- More than 200 million years ago, the Triassic, the first period of the Mesozoic era, ended with the fourth mass extinction. Quebec's Lake Manicuagan, a sixty-mile-wide crater in the northeastern part of the province, was formed at about this time and may be related to that extinction. Did this event, which reduced the world's population of competing animals, actually encourage the development of the dinosaurs? We know very little about that episode from long ago. Since Manicuagan's size seems a little small to account for a major extinction, possibly a series of impacts from a comet shower lasting hundreds of thousands of years might have triggered this extinction.

- There was a smaller extinction of species in the middle of the Cretaceous period, about 91 million years ago. This event took place some 26 million years before the great dinosaur extinction. Its cause might have been an extended period of volcanic activity. Or was there an impact during this time? If the extinction was caused by a comet, the timing would work well with our next subject.

ARE EXTINCTIONS PERIODIC?

With only five or six mass extinctions known since the beginning of life, the idea that extinctions of species can recur periodically seems improbable. But less severe rises in the species extinction rate have occurred at least twenty-four times in the last half-billion years. In 1984, D. Raup and J. J. Sepkoski suggested that extinctions take place on our planet every 26 million years on average. Other researchers have confirmed that some periodicity seems to exist, anywhere between 26 million and 31 million years. Although it is very difficult to confirm this cycle, it may be genuine.[7] If it is, a periodic extinction, caused by comet strikes, is a scary thought to consider. Not only are we assured that some shower of comets is on its way, but we also even know when.

The Earth does not have any periods or cycles that are 26 million years long. The explanation for the cycle, if it exists, must come from the vast movements of the solar system through the galaxy. One possibility is that the Sun has a faint companion star, a red dwarf second sun that loops around in a wide orbit, moving far out into the galaxy, then returning to approximately the distance of the Oort cloud. The star completes an orbit in about 26 million years. As it rips through the Oort cloud, it would send a shower of comets toward us.

There are two problems with this idea. First, there is no other evidence of any sort for such a stellar companion to the Sun, and, second, a star in such a wide orbit could not possibly stay in orbit for more than a few revolutions before some other distant star yanks it away forever.

There may be a better explanation. Our planet moves in several ways. It rotates once in a little under twenty-four hours, revolves about the Sun once a year, wobbles like a top every 22,000 years, and, along with the Sun and the other planets, rushes around the center of the galaxy every 225 million years. Nothing in those statistics offers what we're looking for. But the galaxy of which we're a part is a fairly flat disk in space. As it moves around the galaxy, the Sun has one more type of motion. It moves away from the disk, then goes back through it, and then swings away on the other side and back toward the disk plane. How often does the Sun sweep through the densest part of the plane of our galaxy? About every 30 million years.

Is it possible that going through what we call the galactic plane sets off a gravitational tug that upsets the comets in the Kuiper belt and the Oort cloud, sending more of them into the inner part of the solar system?

I leave this as a question, not as an answer. Like the perplexed mammal at this chapter's start, we look around us and wonder—not at what just happened, but at the vast beauty and immensity of space, and what surprises it might hold in store for us.

CHAPTER SEVEN

EARTH'S FIRST COLD WAR

To see the World in a Grain of Sand,
And a Heaven in a Wild Flower
Hold Infinity in the palm of your hand
And Eternity in an hour.

—WILLIAM BLAKE, "AUGURIES OF INNOCENCE," 1803[1]

T HIRTY MILLION YEARS AFTER the loss of the dinosaurs, an episode occurred that might have been a geologic forerunner of the U.S.-Soviet Cold War. In the late Eocene period, some 35.5 million years ago, at least two big explosions might have led to the loss of many species of life. (The timing would be roughly in tune with the periodical nature of mass extinctions we discussed in chapter 6.) On the United States side, a fifty-mile-wide crater near Washington, D.C., marks the place where a five-mile-wide comet or asteroid came to stay. The crater lies under Chesapeake Bay. The Russian counterpart is a sixty-mile-wide crater near Popigai, east of the Ural Mountains in northern Siberia. Each of these events would have resulted in a global cloud of dust and a long, cold period called impact winter. Such a months-long, worldwide winter would be similar to the much-feared nuclear winter that so many scientists predicted as a result of nuclear war between the United States and the Soviet Union, a winter in which the living would envy the dead. The ancient impacts would have lacked the radiation that follows a nuclear war, so Cold War number one might have been preferable to what might have happened had Cold War number two turned hot.

MEANWHILE, BACK ON MARS

Five million years after the Cold War ended, another comet crashed—but this time the victim was Mars, not Earth. It blasted away tons of rocky material. On Earth, the debris would have risen high into the atmosphere and then crashed back down. But thanks to Mars's lesser gravity, much of the material would have left the planet entirely to become temporary new members of the Sun's family. After journeying through the solar system for 15 million years, at least one of the rocks got too close to Earth, partially vaporized in the atmosphere, and plummeted to Antarctica. This was the rock named ALH 84001, which we met in chapter 4. Surely, it was not the only rock from that impact that changed addresses from Mars to Earth. If one was found, hundreds, if not thousands, probably fell over time from that single impact.

One planetary scientist might have identified the actual launchpad from which ALH 84001 took off. Of the more than forty thousand craters that Nadine Barlow has catalogued from spacecraft images, she notes that two are surrounded by obvious blankets of material ejected from the impact sites. A clearly defined ejecta blanket is a signature of a relatively recent impact. The two suspect craters are also oval in shape; one is fifteen miles long and nine miles wide. These shapes are clues that the impacts were at a sharp angle, which would have helped send the debris on its way into space and eventually to Earth. Moreover, the two craters dot the southern highlands of Mars and are near ancient, long-dried riverbeds. This means that the water that is so important to the interpretation of life did flow nearby at one time.

SEEING THE EVIDENCE

In chapter 2, I suggested that a look at the Moon through a small telescope would reveal a panorama of impact craters, all the evidence to satisfy our curiosity that impacts really did play a role in our past. Thanks to weathering and erosion, the motion of the continents across the ocean floor, and the process of mountain building, Earth has erased most of her craters. There are about 130 known impact structures on the lands of Earth. Most of these are difficult to spot, showing evidence only from the air or as a result of painstaking digging to retrieve samples of rock. Let's explore three of these sites. In 1967, my grandparents and I visited two of them during a cross-continent vacation trip. Barringer Meteor Crater, east of Flagstaff, Arizona, was the first, and far away in Sudbury, Ontario, we drove right through the second. I didn't know about the third one till years later.

BARRINGER METEOR CRATER, WINSLOW, ARIZONA

Signs are not necessary to remind anyone that this is one gigantic hole in the ground. What a sight this is! When I introduced my wife, Wendee, to this crater, I shielded her eyes so that she'd get her first view all at once. "What you are about to see," I told her as we rounded a corner to our first full view, "was dug out fifty thousand years ago, in less than ten seconds." Three-quarters of a mile wide and about two hundred yards deep, the crater is all that's left of an impact of an asteroid made mostly of iron. (Iron meteorites, though uncommon, do fall from time to time.) Had Flagstaff's residents, seventy miles away, been there that long ago, they would have felt the shock. Windows would have been broken, and had they looked toward the east, they would have seen a mushroom-shaped cloud billowing over the impact site. However, they would not have been killed. Many times, news reports of the discovery of an asteroid a few yards in diameter are accompanied by the frazzled warnings that if the rock had hit, it would have destroyed civilization. Unless the asteroid or comet is at least about half a mile wide, that is simply not true.

Although the Arizona crater is officially called Bar-ringer Crater, after the turn-of-the-century explorer who first suspected its impact nature, it is more commonly known as Meteor Crater. The reason is that the U.S. Postal Service once had a station there called Meteor, Arizona, named because of the many dark iron meteorites that litter the surrounding landscape.

The Barringer Crater led the young geologist Eugene M. Shoemaker to explore the role that impacts have played in our solar system. By the end of the 1950s, Gene had determined that this feature was not a collapsed salt dome or the result of a volcanic steam explosion, as others had suggested. His evidence:

- Whole formations of rock were completely over-turned, and were lying upside down. Such violence is not typical of a volcanic eruption.
- Rocks had been melted by shock, then resolidified.
- The structure of the crater was rather similar to two other craters he had studied earlier, craters made not from the sky but from underground nuclear explosions! To Gene's eyes, Meteor Crater was an upscaled version of these.
- In 1960, a mineral called coesite, a type of silica that only the high temperature and pressure of an im-pact could form, was found in the rocks on the crater walls.

Gene's success in showing for the first time that a crater could have an impact origin was the start of a brilliant career. Although Arizona's meteor crater was one of Earth's best examples of an impact feature, Gene believed that impacts were the rule, not the exception, in the story of the solar system. Convinced that most of the lunar craters were the results of impacts, Gene produced the first geologic map of a lunar crater—he chose Copernicus, one of the Moon's largest craters. He trained the *Apollo* astronauts to do geological fieldwork on the Moon. His career was capped with his role in the discovery of Comet Shoemaker-Levy 9, the comet that collided with Jupiter in 1994 and which is the subject of chapter 13.

Dan Durda's original painting of a bright comet over an observatory at twilight captures the sense of wonder that is associated with a bright comet.

THE SUDBURY IMPACT BASIN

The Barringer Crater is an unmistakable blot on the northern Arizona landscape, but the same cannot be said for the crater that begins at the northern edge of the Canadian city of Sudbury, Ontario. Formed 1.9 billion years ago, the crater is at least 125 miles wide. Not only is Sudbury the largest known impact feature on Earth, it also is the second oldest. (Vredefort, in South Africa, clocks in at 1.97 billion years.)[2] As the comet that probably formed it was closing in on the atmosphere, it would have seen an Earth devoid of all but the simplest single-celled life forms. There were no complex creatures to be extinguished by its blast. If there had been, the Sudbury impact would have been at least as devastating as the dinosaur blast over 1.8 billion years later. In the long geologic history of the Earth, however, the two impacts took place long after the late heavy bombardment. To put the Sudbury impact into a time context, a bright comet would have risen majestically over the top of a distant hill as often then as it would appear over the wall of a Sudbury home now. The frequency of comet and asteroid impacts 1.9 billion years ago and 65 million years ago was probably the same as it is now. Both impacts would have topped their "greatest stories of the last hundred million years" feature issues.

I was not impressed with Sudbury when I first saw it in 1967. The nickel mine dominated the city, and great smokestacks belched smoke into the air. When I returned in 1994 to give a lecture, I saw a city transformed. The smokestacks were still there, but the materials they exhaled were now environmentally friendly. Laurentian University's geology department was actively studying the impact basin, and there was a fabulous new science museum and education center called Science North. It's a beautiful place, with an auditorium carved out of ancient rock that has the appearance of having melted and re-formed from the impact.

Sudbury exists because its surrounding rocks are rich with nickel. The impact basin lies just north of its host city. The original crater was much wider and deeper than its present remnant, which was first identified as an impact site by Canadian geologist Robert Dietz in 1963. It is almost a miracle that this crater is visible at all after almost 2 billion years of Earth history. The reason the crater is still there is that Sudbury sits where some of Earth's oldest rocks lie exposed as a large area called the Canadian shield. After Dietz detected the oval shape of the possible crater, he set out to examine the site. He struck pay dirt within ninety minutes! The rock is rich with shatter cones, beautiful conical structures that formed at the moment of impact. With the force of impact, the rock shattered, then melted from the intense heat of crater formation. As each shattered segment of rock solidified again, the lines of stress were left, appearing almost as an arrow pointing to the center of the impact site. Thousands of shatter cones have been found near Sudbury.

Back in 1967, it was also commonly believed that the nickel and other ores that enrich the local rock were direct gifts from the cosmos in the guise of a nickel-rich meteorite. But the nickel was always here, not on the comet. Like a motivational expert bringing out a youngster's hidden talents, the comet concentrated the ores that were already scattered throughout the rock. The force of its impact scavenged out the ore, forcing it around the perimeter of the basin like the shell of an egg.[3]

THE HOLLEFORD CRATER

I was halfway through my master's degree in English at Queen's University in Kingston, Ontario, before I realized that I was studying just twenty miles from a 550-million-year-old crater. A shallow depression is all that's left of it. As a country road works its way through the feature, drivers climb a barely noticeable rise, descend into the crater and, less than a minute later, climb out the other side. Were it not for a plaque identifying the crater, it would be almost impossible to notice.

Although the Holleford Crater is about the same width as the one in Arizona, it is far older and is all that remains of what once was a far larger crater. Unlike the Flagstaff residents of fifty thousand years ago, who would have found the strike next door a mild inconvenience, Queen's University students in Cambrian time would have lost their books, beds, and lives under an explosive concussion of pressure and heat.

Each of the 130 known impact structures on Earth has a different story to tell. Ranging in age from Sudbury's basin to sites younger than Arizona's crater, they are the permanent resting sites of comets and asteroids that all shared the single misfortune of traveling around the Sun in paths that crossed the orbit of the Earth. Such orbits are death sentences. When a comet or an asteroid follows a path like that, sooner or later a collision is inevitable.

CHAPTER EIGHT

A TIME FOR COMETS

With my head exalted I shall touch the stars.

—HORACE, FIRST CENTURY B.C.[1]

Earth has been a busy place these last few billion years. In between comet and asteroid impacts, the building of mountain ranges, the vast shifting of the tectonic plates that support the continents, and the latest scandal in Congress, life somehow had a chance to flourish here.

We left our story about comets and the development of life as our mammalian ancestors were getting larger and more sophisticated. We now rejoin that progress by meeting one of class Mammalia's best representatives, a man who looked out into the sky long ago and thought about comets. His name was Lucius Annaeus Seneca, and he lived in Rome in the first century A.D. His writings made him immortal, but his life was ended at the whim of his boss, Emperor Nero.

Seneca tried to combine the two careers of politics and writing. He fell into this dilemma by going into teaching, his class featuring just one pupil, the young Nero. When Nero became emperor of Rome at the youthful age of seventeen, Seneca found himself to be one of the new leader's most influential advisers. Still indulging his interest in science, Seneca composed *Naturales Quaestiones*, now regarded as one of the most valuable documents about our understanding of Nature ever produced. Seneca's two careers collided head-on in the book's section called "De Cometis." Seneca was well aware that the civilization around him was a cruel combination of the lofty ideals of Nature and the banal events of Roman politics. He tried to keep his head above water by playing one against the other. In "De Cometis," he insisted that comets were formed in our atmosphere from extremely dense air, a view that agreed with the prevailing doctrine of Aristotle.

However, Seneca also believed that once comets were formed, they remained part of the universe forever.[2]

As Seneca was composing his words in the quiet of his study, Emperor Nero was descending into madness. In A.D. 59, Nero murdered his mother, Agrippina, and pressured Seneca to find a reason to make his act somehow acceptable. As Nero's tyranny continued, Seneca tried to stay a step ahead of his murderous boss by recasting the ancient role of comets as harbingers of disaster. Commenting on a parade of comets in the early 60s, in "De Cometis," Seneca insisted that they all were favorable omens to the emperor. It didn't work. In A.D. 65, Nero commanded that his best friend and most trusted adviser be put to death.

The idea that a comet could be considered a warning from angry deities probably stems from actual appearances of comets during the great wars and sieges of our past. Aristotle's theory that comets were "exhalations" in our own atmosphere helped feed this fear. It is a coincidence that one of the most memorable of these comet apparitions, during the Norman Conquest of England in 1066, happened to be an appearance of Halley's comet, the most famous celestial visitor in history. Even though the pace of maturity of cometary knowledge quickened as we approached our century, the old superstitions still survived, and the comets' perceived habit of arriving on our scene just before momentous events probably helped to prolong that superstition. Let's glance at twenty of the best comet appearances in the two thousand years that began with our oldest known record of a comet:[3]

- 1059 B.C.: At the height of a war between two Chinese kings, Wu-Wang and Chou Hsin, a bright comet commanded the sky, its long tail pointing toward the east.

- 974 B.C.: A bright comet appeared near the northernmost part of the sky. Could this have been the same comet that hung like a sword over Jerusalem (see chapter 1) near the end of King David's reign? The timing is reasonably good, but who knows?
- 373–372 B.C.: A magnificent comet coincided with an earthquake and tidal wave at Achaea, Greece. If this was the comet that the Greek scholar Ephorus was referring to as having split into two separate pieces, then this comet was the first of a long comet "tradition" of comet splitting that includes the great comets of A.D. 1106, 1846 (Biela's comet), 1882, 1965, and, most recently, Comet Shoemaker-Levy 9 in 1993.
- 240 B.C.: The first recorded appearance of the comet that would later be named after Halley.
- 162 B.C.: A comet described as a "celestial magnolia tree." The Chinese had some fanciful descriptions for comets, but we didn't know much about them until the famous silk book—a book composed of silk pages—was unearthed from a Han tomb near Mawangdui, China, in 1973. The book shows more than two dozen stylized drawings of types of comets. Composed around the fourth century B.C., the book presents several line sketches of comet shapes, along with interpretations of the type and length of wars, harvests, and other important events they forecast. One of the designs looks troublingly like a swastika, though it is hard to imagine how that hated design could have evolved from an ancient comet.[4]
- 137 B.C.: According to Seneca, this comet's tail spread to "unlimited size."

- 44 B.C.: The comet of Julius Caesar. In Shake-speare's play, Caesar's wife, Calpurnia, tried to keep him from going to the Senate house in Rome, her fear based on a dream. She implored her husband:

When beggars die there are no comets seen,
The heavens themselves blaze forth the death of princes.[5]

We are fortunate that comets in ancient times received such attention, for whenever one appeared, both its track across the heavens and its physical appearance were recorded. Calpurnia might have dreamed of a comet forecasting her husband's assassination, but this comet appeared several months *after* the assassination, during the games that Octavian was holding in Julius Caesar's memory.

- 12 B.C.: This apparition of Halley's comet was suspected by some writers as being the "star" seen by the Magi. The Christmas star is now commonly thought to have been not a comet but a close conjunction of two planets.
- 5 B.C.: A bright comet lasted for more than two months; also a Christmas star candidate.
- A.D. 54: Bright comet in the constellation of Gemini, around the time of Emperor Claudius's death.
- A.D. 59, 60, 61, and 64: Comets cited by Seneca as being evidence of Nero's greatness.
- A.D. 104: A comet had the appearance of "loose cotton."
- A.D. 390: A sign hung "like a column" in the sky.
- A.D. 418: A comet with a tail more than 100 degrees long, trailing across more than half of the visible sky, covered the Big Dipper.

- A.D. 565: A comet lasted more than three months.
- A.D. 837: The closest recorded apparition of Comet Halley to the Earth.
- A.D. 892: Another comet resembling a celestial magnolia tree.
- A.D. 896: A large comet moved eastward with two smaller companions. One observer suggested that they approached each other and then separated as if they were fighting among themselves.
- A.D. 905: A comet described as rapidly changing color from bloodred to a color resembling white silk. Although this color change could have been due to a difference in the clarity of the sky, it would be interesting, indeed, if the change was due to the release of some dust or gas in the comet itself.
- A.D. 1000: A bright comet was reported from France, a country that would later become famous for the many successful comet hunters it produced.

Just as comets continued to appear, watchers continued to see them as omens. Just before Napoleon's invasion of Russia in 1811, a comet appeared. According to a story from a nun named Antonina, the visitation was terrifying:

> One evening, as we were on our way to a commemorative service at the Church of the Decollation de Saint-Jean, I suddenly perceived on the other side of the church what appeared to be a resplendent sheaf of flame. I uttered a cry and nearly let fall the lantern. The Lady Abbess came to me and said, "What art thou doing? What ails thee?" Then she stepped three paces forward, perceived the meteor likewise, and paused a long time to contemplate it. "Matouchka," I asked,

"what star is that?" She replied, "It is not a star, it is a comet." I then asked again, "But what is a comet?" The mother then said, "They are signs in the heavens which God sends before misfortunes." Every night the comet blazed in the heavens, and we all asked ourselves, what misfortunes does it bring?[6]

BRINGING GENERATIONS TOGETHER

It took two comets to bring our understanding out of the realm of superstition. The Great Comet of 1556 inspired Tycho Brahe to propose that Aristotle and Seneca were wrong about comets forming out of thin air. The path that comet took as it crawled across the sky convinced the great Danish astronomer that comets had to be from far out of the atmosphere, even beyond the Moon. More than a century and a half would pass before Edmond Halley proved the extraterrestrial nature of comets beyond any doubt by his brilliant work on the comet that now bears his name.

Halley suspected that comets that appeared in 1531, 1607, and 1682 were returns of the same comet as it meandered its way about the Sun. "Hence I dare venture to foretell," he wrote, "that it will return again in the Year 1758."[7] Halley had as much English pride as he had brains, and he later added, "Wherefore if according to what we have already said it should return again about the year 1758, candid posterity will not refuse to acknowledge that this was first discovered by an Englishman."[8]

Edmond Halley did not live to see the triumph of his prediction. The comet's appearance on Christmas night of 1758, seen by just one German farmer named Georg Palitzsch, was the final proof that this comet was a cousin of the Earth, whirling around the Sun. Not until the

Halley's comet, at lower left, passes by the Milky Way in the spring of 1986. Photograph by Tim B. Hunter.

nineteenth century were celestial mechanicians like J. Russell Hind able to connect earlier apparitions to Halley's comet, and early in this century, Cowell and Crommelin used ancient records to confirm the comet's visits as far back as 240 B.C.

I like to visualize Halley as the great inspector comet. Winging its way from beyond Neptune, the snowball pays its once-in-a-lifetime visits to check on the progress of our civilization. It undoubtedly directed ancient feudal disputes in 240 B.C., and looked on at the last stand of Attila the Hun in A.D. 451. In 837, the comet gave the Earth itself an angry swat as it passed by at close range, its tail covering half the visible heavens. The Norman Conquest of 1066 would not have surprised the great visiting comet, surely by now used to our planet's wars. The comet's return in 1456 saw a hint of change. The Middle Ages were in decline, and civilization was bordering on the new ideas of the Renaissance. However, Pope Calixtus decreed a time of special prayer so that the comet would help repel a Turkish invasion. In 1531, because of royal tradition, England's King Henry VIII should not have seen it during his conflict with Pope Clement VII over his marriage to Ann Boleyn. (Queen Elizabeth broke that tradition by observing the comet of 1582.) Taking a break from writing *Timon of Athens* in 1607, Shakespeare possibly saw the comet. On its next visit in 1682, the comet inspired Halley himself to explore its history and discover that it had been here before. His prediction that it would return in 1758 was a watershed in the history of science. During that visit, the comet oversaw the reigns of France's Louis XV and George II in England with his increasingly restless thirteen colonies. When it next returned in 1835, the United States was its own country under President Andrew Jackson. What history this comet has seen!

THE NINETEENTH CENTURY

Living in the last century would have to be a comet observer's dream. In chapter 1, we shared the fine wines and poetry that accompanied one of the largest comets ever to visit the Earth's vicinity in historical times, the Great Comet of 1811. But what happened on a single night in 1833 might have been even more impressive. On the night of November 17, there was a display of meteors that rivaled any in history. As the Earth passed through the orbit of a small comet, for a few hours meteors fell at the rate of hundreds of thousands per hour. That was the night the Earth collided head-on with the densest part of the comet's dust trail. At its following return in 1866, two amateur astronomers, William Tempel and Horace Tuttle, discovered the comet, and its connection to the meteors was quickly realized.

The mighty comet of 1843 grew a tail that was as long as the distance between the Sun and Mars. Stretching halfway across the sky, this comet was a sight indeed. In 1858, Donati's comet hung in the evening sky with two long tails, one of gas and the other of dust. The comet of 1861 brightened rapidly and steadily after John Tebbutt discovered its faint, blurry light in the sky over his Australian observatory. By the time it appeared over the southern horizon of England, this comet had a nucleus as bright as a first-magnitude star, and grew a protracted tail stretching over most of the sky.

William Ellis, the observer assigned to England's Greenwich Observatory that night, saw the comet rise over the southern horizon. Torn between the work he had been assigned to do and the work he wanted to do, Ellis turned his telescope to the comet in secret.[9] A year later, a comet called Swift-Tuttle appeared with a rare spike directed toward the Sun, and a series of jets erupted from the comet's center. This comet later became known as the parent of the annual Perseid meteor shower. Since his other comet, Tempel-Tuttle, sires the Leonid meteors, Horace Tuttle is credited with discovering the comets responsible for two of our strongest meteor showers.

In 1877, Coggia's comet grew a long tail that stretched for some 45 degrees, hanging, as poet Gerard Manley Hopkins wrote, like a shuttlecock in badminton just as it arcs over its players. Five years later, a great Sun-grazing comet, the Great September Comet of 1882, broke into several pieces after its encounter with the Sun.

THE TWENTIETH CENTURY

The appearance of Halley's comet in 1910 was one of its finest returns ever. As the "Comet Rag" tickled piano ivories, comet pills were advertised to prevent instant death from the comet's supply of cyanogen gas. The discovery of this poison was a big story that year, especially as the Earth went through the outer part of the comet's tail. But the gas was present in far too few molecules to do any harm. That was the year of two great comets. Five months earlier, an unexpected winter visitor, known only as Comet 1910a, brightened snowscapes around the world.

Then came a comet drought. Except for Comet Peltier in 1936 and a bright comet that was found during the November 1948 total eclipse of the Sun, there were no spectacular comets until 1956, when Comet Arend-Roland did a fine performance. The comet featured a bright "antitail" pointing *toward* the Sun instead of away from it. (The antitail appears when the comet crosses the plane of the Earth's orbit and we see sunlight reflected off dust around the nucleus.) A few months later, a second comet, Mrkos, appeared bright in the evening sky.

In 1965, Comet Ikeya-Seki turned my own sights toward comets. It was discovered by two Japanese amateur astronomers that September, Kaoru Ikeya, a worker in a piano factory, and Tsutomu Seki, a guitar instructor. Stung by the failure of his father's business, Ikeya resolved to restore his family name by discovering a comet. Ikeya built a telescope for himself—from grinding the mirror to building the mount. He accomplished this feat for the equivalent of twenty dollars. His ambition was to use his new telescope to find a comet. On the second night of 1963, Ikeya achieved his dream. He realized it again a year later. And on the morning of September 18, 1965, Ikeya achieved his dream a third time. That comet was found the same night by Tsutomu Seki. Ikeya-Seki brightened until, in the third week of October, it was visible in broad daylight as it soared around the Sun. In Ikeya's case, his dream was achieved in a spectacular way.

As a member of a special group of "Sun-grazers," to which the September 1882 comet also belongs, Comet Ikeya-Seki brightened rapidly as it seemed to follow a headlong path to destruction in the Sun. In late October, the comet did a hairpin turn, skipping over the surface of the Sun at a distance of less than 200,000 miles. That October, I cycled to the top of Westmount Mountain near my home to see a surreal searchlight beam—the tail of Comet Ikeya-Seki—climbing out of the St. Lawrence River. Looking at that tail, I was hooked by the magic of how an amateur astronomer, armed with little besides a telescope and perseverance, could find a new world in the heavens.

CHAPTER NINE

A FIELD OF DREAMS

*Time has not lessened the age-old allure of the comets. In some
ways their mystery has only deepened with the years. At each
return a comet brings with it the questions which were asked
when it was here before, and as it rounds the sun and backs
away toward the long, slow night of its aphelion, it leaves
behind with us those questions, still unanswered.*

*To hunt a speck of moving haze may seem a strange
pursuit, but even though we fail the search is still rewarding,
for in no better way can we come face to face, night after night,
with such a wealth of riches as old Croesus never dreamed of.*

—LESLIE C. PELTIER, 1965[1]

WHEN THE TELEPHONE RANG at 5:30 A.M. on a chilly March morning in 1997 for a scheduled interview about Comet Hale-Bopp, I was already out in my observatory, eye at the telescope eyepiece, searching the sky to the east for possible new comets. The observatory is in the backyard of my desert home in Vail, just southeast of Tucson, Arizona. The sky was clear and crisp as I unlatched the metal and wood roof. With a gentle shove, I pushed the roof on its tracks so that it revealed the night sky. Mars was overhead, shining like a reddish beacon in the night, and Comet Hale-Bopp hung like a bridal veil in the northeast. But I had to answer the phone. At the other end of the line was a talk-show host. I was expecting to spend the next half hour talking about the beauty of Hale-Bopp and the majesty of comets, all the while staring at the sky.

I was wrong. "What do you think of thirty-nine people killing themselves," the host asked me, "because they thought a UFO was following the comet?" I had no idea what he was talking about. I was aware that an amateur astronomer in Houston with a lot of equipment and little sense had taken an image of the comet a few months earlier, and that he had misidentified a nearby star and called a radio station to report that a UFO had launched itself from the comet. I explained to my radio host that comets used to incite fear; and that they were seen as bad omens. Unaware of the unfolding tragedy in San Diego, I went on to say that there was but one documented case of a comet causing the death of a person.

The guilty comet, I explained, appeared early in the fifteenth century. John Galeas Visconti, prince of Milan, had asked his astrologers if he had anything to fear, and they replied that he would live without discomfort as long as no comet appeared. The prophecy satisfied the prince, at least until the great comet appeared in February 1402. By the middle of March, the comet grew large and very bright. For a seven-day period beginning on March 22, the comet was easily visible in broad daylight. No other comet has had that distinction for so long a period of daylight visibility. When the great comet neared its greatest brilliancy, the prince went outdoors, took one look at it, had some kind of attack, and died.

That was in 1402, I said to the talk-show host. It would be astonishing that a comet would have that kind of effect on a person six centuries later. The interview ended, and in the brightening dawn, I closed the observatory and went inside to receive a second phone call—now from a wire service reporter, asking the same question. I had to learn fast about the thirty-nine Heaven's Gate cult members in San Diego who offered up their lives to the UFO behind Comet Hale-Bopp. I was stunned. No longer could I talk with some humor about the hapless prince of Milan. Now thirty-nine more people had died from comets.

Or had they? Actually, Comet Hale-Bopp had not killed these people. Just as a mistaken conviction that comets were bad omens killed the Italian prince, their own beliefs killed them. The ancient fears had one very important consequence: They led to the chronicling of observations that allows us to recall the appearance of comets long gone. No doubt, somebody recorded the passage of Comet Hale-Bopp around 2400 B.C., when the pyramids at Giza were only four hundred years old. (Because of the Earth's slow wobble we call precession, the pyramids were angling not toward Polaris, but to a different pole star, Thuban in Draco.) But, in any event, no records of the comet's last visit have been found.

Comet Hale-Bopp, photographed before dawn on March 17, 1997. This eleven-minute exposure was taken by Roy Bishop.

THE ART OF COMET HUNTING

Why do some people spend so much time searching for comets? I have heard as many reasons as there are comet searchers. One successful hunter insists it was to get his name on a comet, while another wanted to prove that a telescope of a certain size was capable of finding a comet. For me, comet hunting is a field of dreams. I enjoy so many aspects of comets: their effect on human history, their appearance in literature, their ghostly beauty, and the fact that they are so interesting scientifically.

The tradition of naming comets after their discoverers goes back to the time of Charles Messier, the first person to find these objects as a result of a deliberate search. His first find came in 1760, just one year after he helped find Halley's comet on its first return since the great Edmond Halley predicted it. The tradition expanded when Denmark's Frederick VI awarded a gold medal for each discoverer of a new comet. The award was presented until 1848. During the late nineteenth century, the American industrialist H. H. Warner awarded a cash prize of two hundred dollars to an American discoverer of a comet. The prize was a powerful motivation for Edward Emerson Barnard, who paid for his house with money won from his comet finds. "This fact proves the great error," Barnard wrote, "of those scientific men who figure out that a comet is but a flimsy affair after all, infinitely more rare than the breath of the morning air, for here was a strong compact house—albeit a small one—built entirely of them. True, it took several good-sized comets to do it, but it was done, nevertheless."[2]

I find that for a successful comet search, it helps to have the perseverance of an Arctic explorer, the heart of a poet, and the patience of Job. The explorer puts up with wind and bitter cold. But once the telescope is set up, and the eye sees the first field of starlight that the telescope offers, the poet feels a sense of deep relaxation as a sightseeing tour of the sky begins. Will the next field bring an odd double star or a star that is unusually red, a strange nebula, a star cluster, or a field of sprawling spiral galaxies? And Job would not have spent his comet-hunting time convinced that the next area must yield a comet, or his hunt would have been short and disappointing indeed. The "do-or-die" approach is not always compatible with a successful comet search.

There is an interesting statistic that comet hunters typically wait an average of four hundred hours for their first find and an average of two hundred hours for every subsequent one, and over the centuries, that has somehow remained a reasonable average time. However, Don Machholz spent some 1,700 hours searching the skies from his northern California site before he bagged his first comet, and I spent 917 hours before I found mine. Sometimes, a find truly comes by accident. The second and third discoverers of Comet Kobayashi-Berger-Milon were looking for a star cluster called Messier 2 in 1975 when they found an elusive fuzzy spot that turned out to be a new comet. Whatever their reasons, comet hunters focus on a single goal, the discovery of a moving patch of haze. The field of a telescope is truly one of dreams when it contains an unknown comet.

By 1970, I had put in several hundred hours searching for comets. That was the year Comet Bennett graced the predawn sky, its long dust tail climbing over the eastern horizon. That comet appeared in the midst of a long period of tension in the Middle East, known as the War of Attrition. Some Egyptian fighter pilots, seeing the comet rising in the east, thought it was a new Israeli weapon. I recall standing on a brand-new highway, not yet opened to traffic but cutting a clear path through the trees to the east. In the darkness, the road stretched to the end of the world and into the sky, where the comet met it like the Emerald City.

Three years later, I could not believe the news reports of Comet Kohoutek. Its New Year's passage around the Sun looked promising enough for one astronomer to predict it would be the comet of the century. But this comet fooled everyone. At its discovery, it was near Jupiter's orbit. It was bright for a comet that far out because it was visiting the Sun for the first time in its more than 4 billion years of existence. As it felt the warmth of the Sun, it expelled some of its primordial material. As the comet approached the Sun, however, it did not brighten as expected. With the science world anticipating a spectacular show, a worldwide cult calling itself the Children of God believed that the comet foretold the end of the world. Cult members stood at street corners in many cities, distributing pamphlets and other literature.

As Comet Kohoutek rounded the Sun just after New Year's Day in 1974, astronauts aboard the Skylab space station were the ones lucky enough to get a fine view. As the comet moved away from the Sun, it faded rapidly, providing Earthbound observers with only a fair show. On one frigid evening, my friend Irving Levi and I drove out a considerable distance from Montreal to spot the comet in a dark western sky. We found a remote site by a highway, and set up my telescope, Minerva. It didn't take long to find the fading comet, its pretty tail pointing upward in the frigid night. As we shivered and looked, our view was interrupted suddenly by the flashing lights of a police car. As the burly cop approached us, he really looked puzzled. We were sure he was convinced we had set up a bazooka. As we explained our purpose in observing the Comet of the Century, the officer obliged us by staring through the eyepiece, one eye shut tight while the other tried to see something, anything, through the eyepiece. We showed him how to focus the telescope, but he looked suspiciously at us, shrugged his shoulders, and sped off into the night. Irv and I continued our observing session for about ten more minutes before the Arctic chill got the better of us.

The Comet of the Century, editorialized the *Montreal Gazette* around that time, slipped by unnoticed. And for the next twenty years, astronomers the world over had to put up with defeated expectations. "Oh, yeah," they'd

hear, "another bright comet is coming? Another Kohoutek?" Actually, the Comet Kohoutek spell should
have vanished with the appearance of Comet West, a
comet whose performance was spectacular in the predawn sky of March 1976. As it rounded the Sun, it broke
apart and presented a fine morning view, which unfor-
tunately was missed by a public poorly informed of this
fine opportunity by a Kohoutek-shy press.

Even Halley's comet did not remove the spell; the
famous visitor gave its worst performance in a thousand
years. When Comet Austin appeared in early 1990 amid
predictions that it would be as bright as Jupiter, no one
seemed surprised that it was barely visible to the naked
eye, and then only from a dark site. It took Comet
Shoemaker-Levy 9's impact with Jupiter in 1994, a spectacle even more profound than expected, to expel the
Kohoutek vexation.

THE NINETIES COMET DECADE

The 1990s will likely be remembered as one of the most
interesting decades for comets in history. The appearance
of four major comets, each different from the others, kept
the subject in the news. But the numbers of comets provided only part of the story. Each comet offered an opportunity for observations by new telescopes, and by new
kinds of detectors attached to telescopes.

The start of the special decade was anticipated by a
conference held in Germany in the spring of 1989. The
idea behind this meeting was to bring scientists together
to discuss the state of what we knew about comets four
years after the 1986 passage of Halley's comet. As the
conference participants headed home, they felt that total

comet understanding had improved vastly thanks to the coordinated effort to observe a single important comet. None of them could have the slightest idea that within ten short years, four of the most interesting comets ever to appear would visit us, one by one, each with its own special story. We explore two of the comets, Levy 1990c and Shoemaker-Levy 9, in other chapters.

THE GREAT COMET OF 1996 . . .

At the end of January 1996, much of the planetary science world was preparing for Comet Hale-Bopp, a big visitor from the outer solar system. Discovered six months earlier, the comet was slowly brightening as it prepared for its first onslaught of the inner solar system in more than four thousand years. But just one person on Earth, a Japanese comet hunter named Yuji Hyakutake (pronounced H'yah-koo-TAH-kee), knew of yet another new comet. Having discovered his first comet only a month earlier, Hyakutake was shocked to find a second new one lurking in the same area of sky as his first.

Any new comet has to be reported to the IAU's Central Bureau for Astronomical Telegrams, where Brian Marsden acts as a sort of celestial cop. Gifted with an encyclopedic mind aided by excellent computer records, Brian heads a committee that assigns discovery credit for new comets and asteroids. When he and his associate, Daniel Green, received a confirmed report of Hyakutake's new comet, they announced it on an IAU circular. Within a few days, more observations allowed them to compute an orbit. The small comet would skip by only 10 million miles from Earth, a celestial hairbreadth, in only five weeks!

The announcement came as a shock. As we'll see in the next chapter, that comet could just as easily have struck the Earth as miss it, and with only five weeks' warning, we would have had as much opportunity to mount a defense as the dinosaurs did. True, the comet's diameter of only two or three miles meant that its impact would not have resulted in a mass extinction, but it was capable of causing major worldwide damage. As the comet approached us, it rapidly headed north and brightened to the point where it could be seen with the unaided eye. One March evening, I stood on the front steps of my friend Tim Hunter's home. As I looked eastward, I could see the fuzzy head and tail of Comet Hyakutake. I took one step to get a better view, and then another. It was one step too many. I fell to the ground and sprained my ankle.

In the nights after my fall, encumbered by cast and crutches, I watched Comet Hyakutake brighten as it pushed its northward course, and its tail grew longer. By the time it reached the area near the Big Dipper, it was the Great Comet of 1996, and the tail was stretching all the way across the sky. For the three nights starting Sunday, March 24, Hyakutake was the most dramatic comet I had ever seen. It wasn't just that it was bright or its tail was long, for other comets, like West in 1976, were brighter. It was that this apparition stayed in the sky all night long, its mighty tail swinging round the north celestial pole. Visible from half the world from dusk to dawn, the Great Comet of 1996 was a treat that I remembered long after the ankle pain was forgotten.

The last Monday in that March was very special. As Hyakutake rose in the sky, it had to fight the bright light of the waxing Moon. But shortly after midnight, the Moon set. By now, the comet was virtually overhead. I traced its mighty tail clear across the sky, from the Dipper south to the constellation of Leo, and on almost to the southern horizon. On the next night, I photographed the comet as it sped by the North Star. Though it had faded a little as it pulled away from Earth, the comet was still a beautiful sight. A bright comet at the celestial pole was a sight I thought I'd never see. On Friday, April 19, my then fiancée, Wendee, and I got our farewell look as the comet sank in the western sky in late twilight. By now, the comet had departed from its graze of Earth and was best seen through our small telescope. "This is just what you expect to see in a comet," she said, "a bright head and a long flowing tail." The comet came and departed quickly, and still another comet was to come.

. . . AND OF 1997

On a clear Saturday night in July 1995, Alan Hale was observing from his home in Cloudcroft, New Mexico. Turning his telescope toward a star cluster first catalogued by comet hunter Charles Messier in the late eighteenth century and called Messier 70, he saw an unexpected faint fuzzy spot. An expert comet observer, Alan suspected instantly that he had snatched a comet. Several hundred miles to the west, a somewhat frazzled Tom Bopp anxiously looked forward to observing with friends. Worried that his creaking, coughing car would not make it to the observing site, he asked his dad, Frank, for his car. Once out at the site, Tom began looking through a seventeen-inch telescope belonging to his friend Jim Stevens. Around midnight, Jim pointed his telescope at Messier 70. As Tom looked through the scope, he noticed the unusual fuzzy spot. While the comet is named Hale-Bopp, the two discoverers spotted the comet almost simultaneously—it is impossible to know who really saw it first.

During the next two years, this comet brightened from a diminutive blob of light to a beautiful ghostly beacon as bright as the sky's brightest stars. While heading our way in April 1996, one month after Comet Hyakutake made its magnificent showing in the sky of Earth, Hale-Bopp passed some 70 million miles from Jupiter, a chance encounter that cut its orbital journey about the Sun in half. As it rounded the Sun a year later, under the clear sky of La Palma, Spain, astronomers at the Sir Isaac Newton telescope discovered a new kind of tail. Some 7.5 million miles long and half a million miles wide, this tail was composed entirely of sodium atoms, something never before detected in any comet.

Observing with the Hubble Space Telescope in September 1996, Harold Weaver discovered the comet was expelling some nine tons of water *each second!* For some months after, the telescope was unable to image the comet during its best viewing phase because it was too close to the Sun. When the big space telescope again pointed toward the comet in late August 1997, it found the comet pouring out water at the same rate as a year earlier, by far the highest rate of water production ever observed in a comet.

Other observers learned something about the comet's origin by *not* finding what they were looking for. Using the *International Ultraviolet Explorer* satellite, observers Michael Mumma and Vladimir Krasnopolsky failed to detect any neon gas in Hale-Bopp. Since neon condenses at only 25 degrees above absolute 0, (more than 400 degrees below 0 Fahrenheit), the two observers expected to find out if the comet had formed in the frigid region of the solar system called the Kuiper belt, the area that includes the planet Pluto. The comet's lack of neon gas suggests that it did not form in the Kuiper belt. Instead, the comet was put together, piece by piece, in a warmer part of the solar system, possibly as close to the Sun as the planet Uranus. As the comet swung around the solar system, it encountered the giant planets. Their gravity forced the comet out into the sphere we now call the Oort cloud. Hale-Bopp probably stayed there for billions of years before starting on a new path that took it back toward the Sun.

Billions of years after its birth, Comet Hale-Bopp swung into the evening sky in the last weeks of March 1997, hanging swordlike in the northwest as comet parties around the world attracted viewers by the millions. I had two favorite views of this comet. One took place in February 1997, on the shore of a frozen lake in southern Ontario at the home of my friends Leo Enright and his wife, Denise. Solidifying from the dense cold, the ice cracked deep beneath us as our group, which included discoverer Tom Bopp, gazed at this celestial wonder. Despite the almost full Moon, the comet was radiant, its tail climbing majestically into the night. My second favorite view was on March 23, 1997. That was the night Wendee and I opened our home and observatory to more than a hundred guests who were there to help celebrate our wedding. The evening began with a view of the comet low in the western evening sky. As the comet set, the Moon rose, and within two hours was in the midst of a deep eclipse. What better way to begin married life than with an eclipse and a comet? For us, it was a field of dreams hoped for and fulfilled.

CHAPTER TEN

TARGET EARTH

I keep picturing all these little kids playing some game in this big field of rye and all. Thousands of little kids, and nobody's around—nobody big, I mean—except me. And I'm standing on the edge of some crazy cliff. What I have to do, I have to catch everybody if they start to go over the cliff—I mean if they're running and they don't look where they're going I have to come out from somewhere and catch *them. That's all I'd do all day. I'd just be the catcher in the rye and all.*

—J. D. SALINGER, *THE CATCHER IN THE RYE*, 1945[1]

WILL A COMET OR an asteroid strike the Earth someday? Yes. When? We do not know. What will be the damage? That depends upon the size and the velocity of the intruding object. There is cause for concern: just because we do not know if we will be involved in a car accident, most people would not drive down their driveways without being covered by auto insurance. We wear seat belts as we drive cars with rearview mirror systems and other devices designed to protect us. We do these things because they are safe, sensible, and appropriate to do.

There are other risks in life. Our chances of perishing in an airline disaster, for example, are about 1 in 20,000. Of the many ways of calculating statistics, one suggests that this is the same as our chance of dying in a comet or asteroid hit. What makes the two types of disasters comparable is the factor of risk versus consequence; an air crash might kill hundreds of people, but a comet hit, which occurs far less often, would affect almost everyone on Earth.

We get plenty of evidence that airplanes can crash, and as a result, governments and industry pour billions of dollars into making flying as safe as possible—far safer than driving a car. But we also get plenty of warnings that dangerous comets and asteroids are around; in fact, every two hundred years or so a comet misses us by less than 3 million miles. How many people are involved around the world in all the efforts to search for asteroids and comets that could pose a threat to life on the Earth? This question is a favorite of David Morrison, director of space at NASA's Ames Research Institute. His answer: Fewer than work a single shift at one McDonald's restaurant.

WHAT THE STATISTICS TELL US

The larger the impacting objects are, the rarer their falls to Earth. Sand-grain-sized particles fall every second; rock-sized particles every hour; TV-set-sized bodies every day. In 1908, in Siberia's Tunguska River area, an asteroid the size of a small building blew up in the atmosphere. Although it did not hit the ground, the shock wave knocked down a forest and killed thousands of deer and other animals. An event of this size, which includes the fall that carved our Arizona's Meteor Crater some fifty thousand years ago, takes place about once in a hundred years.

An object the size of a small village, like a piece of Comet Shoemaker-Levy 9, might hit Earth once every 100,000 years (not often enough for us to live in fear, but often enough for us to be aware). A comet the size of a small city, like the one that probably led to the demise of the dinosaurs, might strike once in 100 million years.

What these statistics mean is not entirely objective. Last year, I had surgery for prostate cancer, and two months later had more surgery for kidney cancer. The tumors were unrelated and both had been found before they spread. However, when my doctor told me that the chances of having the two unrelated tumors in men my age were less than a tenth of a percent, the figure was not comforting. For me, the chances were 100 percent. As I lay in the recovery room, I thought at least I had won out on another statistic, that along with everyone else on Earth, I had beaten the 1 in 100,000 chance that a comet or asteroid would strike the Earth in the same year as my surgery. You can do anything with statistics.

WHAT ARE WE DOING TO SEARCH FOR
DANGEROUS COMETS AND ASTEROIDS?

The only way to beat the odds is to locate physically every possible asteroid or comet that could pose a threat to the Earth. A few years ago, a number of successful programs were under way to do that. Using wide-field cameras, they scanned the sky—virtually, all the sky—every month in search of the elusive comets and asteroids. These programs used old-fashioned film and "Schmidt Camera"–type telescopes to do their work. The Schmidt is a beautiful design that allows large areas of the sky to be photographed at once. On a single film, observers can record faint stars, galaxies, and comets and asteroids over an area covering almost twenty diameters of the Moon. The Schmidt is essentially a very efficient bucket for starlight that allows much of the entire sky to be photographed over several clear nights.

For two decades, from 1973 to 1994, one eighteen-inch-diameter Schmidt telescope, located in the mountains north of the California city of San Diego, was the linchpin of two important programs. The first program, Planet-Crossing Asteroid Survey (PCAS), began in 1973 as the brainchild of Gene Shoemaker. Working with Eleanor Helin, Gene's strategy was to take a series of long-exposure photographs of the sky, recording the motions of fast-moving asteroids as they leave little trails of light on the film. Some six years after the project began, Gene changed its direction. Bringing his experience as a geologist to bear, he developed a stereomicroscope of the type normally used to look at aerial photographs for use in astronomy. By placing two photographs of the same area, taken from very slightly different positions, into a stereomicroscope, the viewer gets an impression of depth. In astronomy, the Shoemaker stereomicroscope looks for depth not in space but in time. The films are exposed using the same telescope at the exact same location, but are shifted in time by some three-quarters of an hour. Someone looking through the instrument sees both films at the same time. The distant stars appear as single images. But any moving object, like an asteroid or a comet, appears not as a trail but as an image raised above the background of stars. What an elegant way of searching!

In 1983, Gene Shoemaker left PCAS to Helin and launched the second program, Palomar Asteroid and Comet Survey (PACS). The new Shoemaker program emphasized major coverage of the entire sky. Thanks to the two competing projects, by 1986, the northern sky was being covered efficiently, with the resulting discovery of a record number of asteroids and comets. In 1987, Henry Holt joined PACS, enabling the program to thrive during the summer months while the Shoemakers were off in the Australian outback studying the resting places of asteroids that had come to Earth to stay. In 1989, Holt and Norman Thomas discovered an asteroid, later numbered 4,581 and named Asclepius, a third of a mile across a few hours after it sped by Earth on March 23 that year, missing our world by 400,000 miles. Had Asclepius hit the Earth that day, the devastation would have been worldwide—not enough to trigger a mass extinction, but certainly enough to disrupt the normal flow of civilization for a year at least. Striking just as springtime began in the Northern Hemisphere, the crash of Asclepius would have sent enough dust into the upper atmosphere to block sunlight and shut down an entire growth season in the hemisphere where most of our food supply exists. Asclepius will probably hit the Earth someday.

A third program, called Spacewatch, is the heart and

soul of Tom Gehrels, one of the most experienced asteroid observers of this century. In 1981, he began his program by merging the old and the new. A thirty-six-inch telescope, built in 1921, was used with a modern electronic imaging chip called a charge-coupled device (CCD). By 1990, after several years of testing and tentative operation, the program was in full production mode, using a state-of-the-art CCD chip. The team discovers an average of two thousand asteroids *per month,* of which a few turn out to be on Earth-crossing orbits.

In the Southern Hemisphere, a fourth program called Anglo-Australian Near-Earth Asteroid Survey has employed the famous 1.2-meter (or forty-eight-inch) Schmidt telescope at Australia's Anglo-Australian Obser-

Total eclipses of the Sun terrified early people into keeping records of celestial events, and might have led to the creation of structures like Stonehenge, and eventually to modern observatories. This image was taken in the middle of a spring afternoon, during the 1970 total eclipse of the Sun. The dark landscape and dark cloud cover turn day into night. Photo by David H. Levy.

vatory. Under a constant threat of financial cutoff, nevertheless, the program has encountered a considerable success in discovering asteroids too far south to be seen by the Northern Hemisphere programs.

Despite the favorable reactions these programs have received, by the end of 1994, PCAS and PACS had shut down, and the southern program finally lost its funding at the end of 1996. Of these original programs, only Spacewatch was still thriving in 1997. On the horizon are three interesting new ventures. First, Spacewatch is expanding to a seventy-two-inch-diameter telescope, one that should increase their discovery rate by a factor of 10. Second, using a satellite-tracking telescope called GEODSS, for Groundbased Electro Optical Deep Space Surveillance System, the U.S. Air Force has combined efforts with a team of asteroid experts at NASA's Jet Propulsion Laboratory, including Eleanor Helin, to search for asteroids near the Earth.[2] The effort has already discovered two comets, each named NEAT after the program's acronym Near Earth Asteroid Tracking program. Third, in Flagstaff, Arizona, Lowell Observatory Near-Earth Object Search (LONEOS) is ramping up to full operation. Using a Schmidt camera coupled with an array of CCDs, the program is expected to cover much of the sky each month. Another program, called LINEAR, is sponsored by the Massachusetts Institute of Technology's Lincoln Laboratory.

Most solar system experts would like to see a more ambitious search program. Called Spaceguard, it would combine the existing programs with newer telescopes spaced strategically around the globe. Instead of discovering all two thousand Earth-crossing asteroids within a century, a probable result of maintaining the current search rate, with Spaceguard we would know within a decade if a sizable asteroid was headed our way. We'll discuss our future more in the final chapter.

WHAT IT'S LIKE TO BE PART OF THE SEARCH

I joined PACS in late 1988, and learned quickly how intensive, and how invigorating, the search for "Earth crossers" could be. Being a part of this program is like being a night watchman for the planet. What's out there with our number on it? It is a challenge I wanted to face—hoping that someone, in one of these programs, would find a threatening intruder before it was too late. I've also observed as a guest on some of the other projects, especially Spacewatch. Although the details of the observing programs, procedures, and demands vary from place to place, the sense of purpose is the same. For each PACS observing week, the Shoemakers and I drove some five hundred miles from our Arizona homes to Palomar Observatory. Crossing deserts and mountains, we tried to complete the drive in about eight hours. Gene and Carolyn had the better time doing it; stopping off in Prescott for an hour's dinner and dancing, they completed the drive by 3:00 A.M. The idea was to use the road to turn day into night, thus altering our biological clocks. By the time we arrived at our home away from home at Palomar, we were ready for hours of sleep that would help keep us going for the taxing all-night observing sessions that were to come.

The best part of my observing session is the start of its first night. With a sky so blue that you could reach out and touch it, the night drops like a curtain on a stage. The stars come out one by one, as they did for ancient Chaldean shepherds watching their flock five thousand years ago. Looking up at the stars, there is little to say what millennium it is.

Our observing procedure was crafted for maximum efficiency. Needing to photograph each area of the sky approximately forty-five minutes apart, we scheduled four fields of sky for each set. After I loaded a sheet of film into the telescope, Gene Shoemaker read to me the celestial position of the star I would be following for the upcoming exposure. We set the telescope, centered the star in the eyepiece, and opened the telescope's shutter. For the next eight minutes, the telescope and I played a sort of celestial video game with the star as referee. The object, to keep the star precisely centered, was thwarted each time the telescope's motor drive made the telescope lurch to the east or west. I corrected these wanderings with a quick push of a button. At the end of the exposure, we replaced the film, moved the telescope to the next area of sky, opened the shutter again, and exposed the film. After the fourth exposure, we repeated the shots. This way, we had two films of each area of sky we chose, films that might hold the precious gold of a new asteroid crossing the Earth's orbit, or a new comet.

As the dawn interrupted our headlong race for data, we had taken and developed between fifty and sixty individual photographs. Did we have a discovery? We wouldn't know until Carolyn scanned these films a day or two later. In all likelihood, nothing unusual disturbed the peace of our films. But in more than twenty-six thousand pieces of film shot over more than two decades, thirty-two comets using the Shoemaker name were detected, as well as about forty-five asteroids that could someday threaten Earth.

On the same night, other programs were scanning the sky. We were in good contact with Spacewatch, and once or twice relied on their superior telescope firepower to confirm a very faint object we discovered. Between our skill in covering large amounts of sky, and Spacewatch's technical ability to penetrate deeper into the sky than we

Ancient comet watchers might have used England's Stonehenge as an astronomical observatory. Construction began around 1800 B.C. Photograph by David H. Levy.

could, both programs flourished over many years without scooping the other's finds. By the time PACS stopped at the end of 1994, I was proud to have played a role in the discovery of some of the comets and 225 asteroids now known to present a threat to the Earth. The orbits of some of these asteroids will change over time.

GRAZING COMETS

If the Earth hasn't been hit in the past two centuries, has it *almost* been struck? Quite often, actually: let's explore the close calls our planet has had.

In June 1770, a comet called Lexell tore by at a distance of 1.4 million miles, a vanishingly small distance in the vastness of space. A little over ninety years later, the Earth passed through the tail of the Great Comet of 1861 as the nucleus stormed past at a distance of 10 million miles. In 1910, the Earth passed through the tail of Halley's comet. As we've seen already, that visit terrified some people into believing that the tail's supply of cyanogen gas would immolate everyone on Earth.

The next close shave did not come until 1983, when an Earth-orbiting spacecraft called *IRAS* (for *Infrared Astronomical Satellite*) and two amateur astronomers discovered a fairly bright comet. Although the *IRAS* discovery came first, slow reporting to Brian Marsden's Central Bureau for Astronomical Telegrams meant that the comet was not announced until only two weeks before its closest approach to Earth. In May 1983, this comet whizzed by the Earth at a distance of less than 3 million miles. On Monday evening, May 9, 1983, I invited a group of friends for a comet party in our backyard. Sitting on lawn chairs and peering through binoculars, we watched fascinated as the comet glided slowly through the sky. The view through a telescope was spectacular. In order to follow the comet, I had to keep the telescope physically moving all the time! This comet was truly having a near miss with Earth.

A PROGRAM TO DEFLECT AN INCOMING
ASTEROID OR COMET

Since comets are rarely found more than a year before they reach the vicinity of the Sun and the Earth, I cannot see us mounting a defense against an attacking comet. Asteroids are different, though. Earth-threatening asteroids orbit in paths that keep them always in the inner part of the solar system, passing close to Earth every few years. It is possible to discover virtually all the larger objects well in advance of impact, giving us decades to prepare for it. With this much warning, we would have the time to implement the following program to deflect an incoming asteroid.

Suppose that an asteroid has been found with an orbit that comes close to Earth several times in the next thirty years, and that the asteroid could strike the Earth at that time. The first step is to mount an intense observation campaign to improve our knowledge of the object's orbit, including what effects successive passes near Earth would have on it. If a collision seems even more likely after that, step two involves launching a reconnaissance spacecraft to study the asteroid, determine whether it is made of soft stone or solid iron, and plant a transponder on its surface. Signals from that device would then allow us to track the asteroid accurately as it soars through space, permitting us to calculate not only the hour of impact, but also the place on Earth, with an accuracy of less than a mile.

The next step, moving the asteroid, is most efficiently done at the moment it passes the perihelion point of its orbit, the place in its path closest to the Sun. At that moment, the vaporization of rock from a nuclear warhead exploding near the asteroid would change the path by a small amount. After the blast, we would measure the asteroid's change in velocity and the new orbit that results. A few years later, when the asteroid next reaches perihelion, a second shot would add to the change. By now, the asteroid should be in a new orbit that would miss the Earth. Doomsday thus averted, we would continue to monitor the object closely to see what its future path will be.

Other, nonnuclear, approaches have been suggested. One interesting possibility is to use the asteroid's own mass to push it away. This "mass driver" approach would work especially well for comets, whose dust could easily be used for fuel. The idea is to attach rocket motors to the surface and then use the object's own substance as fuel.

Four thousand years after Stonehenge, a small telescope and the mighty four-meter telescope dome pose at Kitt Peak National Observatory. Photograph by David H. Levy.

Over a period of time, this braking or acceleration should change the orbit enough to turn a catastrophic collision into a near miss.

Whatever plan we use, before we try to deflect a body in space, we have to know it exists. That's why the current search programs, and the worldwide effort called Spaceguard, are so important. To help coordinate this effort, the Spaceguard Foundation was formed recently. Headquartered in Rome, the organization's purpose is to bring together scientists in diverse fields that are related to cosmic impacts, and to focus public attention on what could be a serious threat to civilization.

A LOOK INTO THE FUTURE

As we've already seen, Comet Hyakutake's close visit in 1996 was truly spectacular. What awaits us in the future? We do know that Finlay's comet will pass within 4 million miles of Earth in 2060, and Comet Giacobini-Zinner will make a close approach in 2112.[3] Twenty-two years later, Halley's comet will brush by at a distance of 7 million miles. But what of the virtually unlimited number of undiscovered comets?

What about the asteroids? As we scan a list of these bodies making passes of Earth in the next three centuries, some names ominously appear again and again. The asteroid Toutatis, for example, makes a close approach in 2004, and another in 2069. Another asteroid, Hathor, tells a similar tale. It is an Aten-type asteroid, which means that its orbit carries it as close to the Sun as Venus. It walks by Earth in 2045, passes us again in 2069, in 2086, and once more in 2130.[4] Its 2086 visit, in fact, will be the closest of any known asteroid, tearing by at only twice the distance of the Moon. We know that before the year 2200 Hathor won't collide, but *someday it probably will.* It is almost as if that little world comes by from time to time, quietly, with a haunting reminder that it is always out there, and that some fateful day it will meet us.

Hathor will make the closest approach of any *known* asteroid. There are an estimated two thousand Earth-crossing asteroids at least half a mile wide that have not yet been discovered. I can write nothing about any of these objects. One could be poised to strike next week. If someday we knew the whereabouts of each of these objects, we would know the risk and, as we have seen, might even be able to do something about it. Is our destiny, then, to be Earth's "catcher in the rye," to find these little intruders before they strike the Earth?

Also long after Stonehenge, other astronomers discovered almost fifty comets from this observatory at Palomar Mountain. Photograph by Jean Mueller.

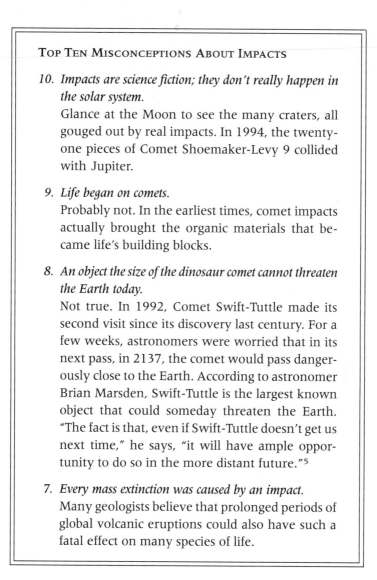

TOP TEN MISCONCEPTIONS ABOUT IMPACTS

10. *Impacts are science fiction; they don't really happen in the solar system.*

    Glance at the Moon to see the many craters, all gouged out by real impacts. In 1994, the twenty-one pieces of Comet Shoemaker-Levy 9 collided with Jupiter.

9. *Life began on comets.*

    Probably not. In the earliest times, comet impacts actually brought the organic materials that became life's building blocks.

8. *An object the size of the dinosaur comet cannot threaten the Earth today.*

    Not true. In 1992, Comet Swift-Tuttle made its second visit since its discovery last century. For a few weeks, astronomers were worried that in its next pass, in 2137, the comet would pass dangerously close to the Earth. According to astronomer Brian Marsden, Swift-Tuttle is the largest known object that could someday threaten the Earth. "The fact is that, even if Swift-Tuttle doesn't get us next time," he says, "it will have ample opportunity to do so in the more distant future."[5]

7. *Every mass extinction was caused by an impact.*

    Many geologists believe that prolonged periods of global volcanic eruptions could also have such a fatal effect on many species of life.

6. *Impacts are bad for life.*
Without impacts, life would lack the diversity we see today. Impacts upset the apple cart of evolution, allowing a burst of speciation and new life forms to prosper.

5. *The Earth is just as much at risk now as it was in the past.*
No. During the period of late heavy bombardment some 3.9 billion years ago, dozens of bright comets graced the sky at any one time, and impacts were far more frequent.

4. *The chance that a comet or asteroid that could damage the Earth's ecosystem will land in our lifetime is virtually zero.*
Not true. According to several studies, not the least of which was completed by Gene Shoemaker, the chances are that a half-mile-diameter asteroid or comet would strike the Earth once in 100,000 years. Statistically, that means there is a 1 in 1,000 chance that such an object will hit during a human lifetime.

3. *To prevent an impact, we have to destroy the comet or asteroid.*
No. A much more favorable option involves detonating a weapon off the bow of the object. The vaporization of material from the intruder's surface would then gently shove it into a safer path.

2. *Any object that hits the Earth could cause global devastation.*

   Not true. Small meteorites fall almost weekly on some portion of Earth without causing any damage at all. For instance, on March 8, 1976, the fragments of a large meteorite fell over Kirin, China. Of the more than one hundred fragments that fell, one weighed in at almost two tons.[6] No real damage resulted.

1. *It can't happen to us. Things won't change after a major impact.*

   It can and, sometime in Earth's future, it will.

CHAPTER ELEVEN

THREE LITTLE WORLDS

Hey Jude, don't make it bad,
Take a sad song and make it better.

—THE BEATLES, "HEY JUDE," 1968

THREE WORLDS—EARTH, VENUS, AND Mars—are lit by the same Sun, and see the same sky. Earth is the home of an indescribable diversity of life. Plants thrive in the sea and on the mountaintops, while animals beyond number swim, crawl, walk, run, and fly. Venus and Mars are desolate. Birds do not soar through Martian skies, and tall mammals do not stroll the surface of Venus.

What happened to Venus and Mars? The answer to this profound question not only defines why these planets failed to support life, but also shows how fragile our own planet is. When comets delivered the building blocks to Earth billions of years ago, they also attempted delivery on Earth's sister planets. It is useful to explore how life failed there.

For the first billion years of the history of the solar system, the Sun was cooler than it is now, and the atmospheres of Venus, Earth, and Mars were remarkably similar. Rich in carbon dioxide, the three atmospheres also contained water vapor that should have formed liquid oceans, even on primordial Venus. In the late 1970s, spacecraft detected an isotope of hydrogen, called deuterium, high in Venus's atmosphere. Venus has about 150 times the amount of this "heavy hydrogen" that Earth has, suggesting that over time, Venus lost most of its normal hydrogen. Where was this normal hydrogen? It probably was in molecules of water that split apart as the Sun grew hotter. It is possible, then, in that brief early period, life might have gotten a foothold on Venus, only to be exterminated as Venus's greenhouse grew hotter.[1]

We move forward 3.5 billion years. The three planets now have atmospheres startlingly different from one another. Mars's atmosphere is very thin, and Venus's air is incredibly thick. We have already seen that, at best, Mars is home to single-celled submicroscopic life, and Venus is probably sterile. What happened?

THE GREENHOUSE EFFECT ON VENUS

Imagine a greenhouse, built to keep a collection of plants healthy and beautiful. On a sunny day, the ground and the plants are heated. The plants radiate the heat back, but not at the same wavelength that it comes in. The outgoing radiation is at infrared wavelengths, which the greenhouse glass will not let through entirely. As a consequence, the temperature inside the greenhouse warms up.

Now suppose our greenhouse is made of glass so opaque to infrared radiation that *no* radiation can escape. Then the temperature inside the greenhouse would continue rising indefinitely. That's Venus. The planet's greenhouse is made not of glass but of a dense atmosphere rich in carbon dioxide and sulfuric acid. The idea of a greenhouse effect on Venus was first suggested by Carl Sagan.[2]

A planet much like the Earth in size, Venus is catastrophically different from Earth. The victim of a runaway greenhouse effect, Venus's problem began because it receives about twice as much sunlight as Earth, which raised the planet's surface temperature. As a result, the ocean water began to evaporate faster, increasing the amount of water vapor in the atmosphere. The water vapor increased the efficiency of the greenhouse effect, which produced more carbon dioxide, which again raised the surface temperature. Eventually, all the water in the planet's oceans evaporated, and the surface temperature soared to hundreds of degrees. Meanwhile, the Sun's ultraviolet rays, much stronger on Venus than on Earth, worked with other chemical reactions to rid the atmosphere of most of its water vapor.

Mariner 2, the first spacecraft to reach another planet, passed by Venus in December 1962 and recorded surface temperatures of more than 800 degrees Fahrenheit. Venus's temperature is high enough to fry almost anything. When I tell children about Venus, I ask them to imagine a Venusian creature shivering as it arrives in our classroom, shivering even as we put it into an oven set on broil. The temperature needs to be almost twice as high as the broil to warm the Venusian to a comfortable level. Because of the dense atmosphere, the air pressure there is about a hundred times higher than on Earth, and its temperature some ten times hotter. The clouds let less than 3 percent of sunlight reach the planet's surface. Thanks to these clouds, the surface is dark by day, but at night, the heat causes the ground to glow.

WHY DOESN'T MARS HAVE INTELLIGENT LIFE?

Although we now know that Mars does not have any higher forms of life, the romance of the planet leaves us to wonder why. Mars revolves about the Sun in a little less than two Earth years. Its day (called a "sol") is only forty minutes longer than Earth's day. When *Mars Pathfinder* landed on a windy plain called Ares Vallis on July 4, 1997, its tiny six-wheeled rover drove through very thin, dust-laden air. The best the summertime heat could climb to was a frigid though livable 15 degrees Fahrenheit, but nighttime lows plunged to minus 110 degrees Fahrenheit.

While these conditions are harsh, they are not completely forbidding for the evolution of life. Earth's poles are almost as cold, and they host many species of life. However, the unbelievable cold of Mars is accompanied by a complete lack of water, on the ground or in the air. *Pathfinder's* home in Ares Vallis is the remains of a flood that took place billions of years ago. Mars was wet then, but over time, the water evaporated into space, and Mars's cold surface took much of the carbon dioxide out of the atmosphere. Much of the carbon dioxide now resides in the rocky soil beneath the planet's surface.

Among other reasons, life seems to have flourished on Earth thanks to an accident of placement. Venus and Mars had the misfortune to form at the wrong distances from the Sun. It seems likely that life did get a foothold on Mars and might even still exist there today in submicroscopic form. Because the planet's gravity was not strong enough, some of its fledgling atmosphere was lost into space. Much of what remained of the air was absorbed into the ground because the planet's temperature was not high enough to keep it as air. In any event, any evolutionary attempt at more complex life forms was doomed.

The argument that Mars lost its atmosphere because its gravity was too weak to maintain one seems to make sense until we explore other worlds in the solar system. Titan, Saturn's largest moon, is almost as large as Mars; its diameter of 3,200 miles is not much less than Mars's 4,200. However, Titan has an atmosphere so dense that we have never seen the moon's surface. Why does it have such a dense atmosphere? The answer probably has to do with Titan's birth. Its orbit indicates that it was not a captured runaway body, or formed as a result of a collision like our Moon, but that it was created at the same time and of the same raw materials as Saturn itself. So, like Saturn, Titan has a thick atmosphere.

INTELLIGENT LIFE IN THE IMAGINATION

Whenever we think of intelligent life, we are drawn to the red planet that lurks so often in the night. In 1938, both amateur and professional astronomers were recording color changes on the planet that seemed to indicate changes in vegetation with the seasons. By the end of the 1930s, most scientists accepted the idea, since proved wrong, that plant life inhabited Mars. Meantime, the question of whether *intelligent* life thrived on Mars continued to be hotly debated.

On the night of October 30, 1938, listeners to the CBS radio network were stunned by what appeared to be an interruption in music programming. A report had been received, the announcer claimed, of some kind of crisis happening in New Jersey. The music returned briefly, but the announcer broke in again with the news that an invasion force from Mars had landed in New Jersey. Of the program's rapidly growing audience, one listener was Patsy Tombaugh. "Clyde," she called to her husband, "there's something about a landing from Mars." One of the most famous astronomical observers in the world, Clyde Tombaugh had discovered the planet Pluto only eight years earlier. His lifelong fascination with Mars quickly surfaced when he heard of the landing. Turning up the volume of the radio, he and Patsy listened to the ongoing reports of utter devastation taking place in New Jersey. Clyde's tentative reaction was that now we know for certain, there is life on Mars.

Other than having symbols of Mars painted on their ships, though, how were we to know that the invasion force was from the red planet? Observations from a Mount Jennings Observatory, the announcer explained,

Venus, the Earth's sister world, orbits closer to the Sun than Earth; Jupiter orbits at a much greater distance. In this photograph by Roy Bishop, Venus is on the left, Jupiter on the right.

detected a bright flash of light just two days earlier that must have been from the launch. Clyde rose from his chair disgusted. He had never heard of Mount Jennings Observatory. And Mars, he knew, was at that time in line with the Sun and unobservable from Earth!

If even Clyde Tombaugh was almost fooled, one can imagine the panic that the CBS presentation set off. It was really an episode of Orson Welles's *Fireside Theater;* their play that night was H. G. Wells's *War of the Worlds*. The only real launch that took place that week was Orson Welles's career. The fictitious Martian invasion led to real riots in several cities as people feared for their lives.

No matter which direction the facts of observation turn, many people still think of Martians as a real people. The changing colors that visual observers note are not caused by vegetation but by clouds and other subtle changes. If the tiny life forms on the Mars meteorite (see chapter 4) turn out to be real, then, in an abstract sense, the age-old excitement about life on Mars would have been justified. There are Martians, but these creatures are submicroscopic, single-celled organisms. These creatures lived, and may even still live, on a world that looks hauntingly like parts of the Earth. Someone taking a cursory look at a *Viking* or *Pathfinder* photograph could mistake their subjects for a California desert. The two landscapes are quite similar. However, the California desert never freezes to −110 degrees Fahrenheit. Mars's polar icecaps are not water but carbon dioxide, and dust storms, such as the one that the spacecraft *Mariner 9* witnessed in 1971, blanket the planet for months on end, preventing much sunlight from reaching the planet's surface.

Mars lacks another process that helps limit the size of mountains and valleys. The Earth's continents ride on tectonic plates that slowly twist and travel across the planet. Hawaii offers a good example of this process. A gateway from the Earth's mantle opened long ago, spewing lava into the sea. As the tectonic plate shifted above the opening, mountain after mountain was formed. The southern end of Hawaii's big island is now over that opening. On Mars, Olympus Mons has been sitting atop a similar opening for billions of years, allowing a single mountain the size of Arizona to grow. Thanks to the absence of tectonic plate movement, Martian mountains are much higher, and the canyons are far deeper.

HOW CAN MARS AND VENUS HELP US?

Although our sister worlds have turned out so differently from ours, they can teach us about our own. Although we can learn something of these worlds by examining them with telescopes, by far the best way is to study them close-up with small robotic spacecraft. Elegant ambassadors from the people of Earth, these craft have allowed us to visit every planet in the solar system except Pluto, and have returned a bonanza of information about surface and atmosphere. Of the many lessons the other worlds have taught us through these craft, one of the most important is weather forecasting, still an inaccurate science thanks to the almost unlimited variables of our atmosphere. A falcon lazily turning circles in the sky, for instance, can conceivably start a whirlwind of low pressure that grows into a great storm within a few days. How air circulates in our atmosphere is a very complex question, difficult to answer because our atmosphere is the only laboratory we have. Or is it? It takes a well-recognized expert in planetary atmospheres like Clark Chapman to point out that studying other worlds helps us to understand our own:

> The atmosphere of Venus is much thicker than ours, that of Mars much thinner. The air on Mars is warmed directly by the sunlit ground (except during planet wide dust storms), while Jupiter's clouds are warmed both from the outside by the Sun and from the inside by the giant planet's own internal heat. Venus spins much more slowly than the Earth, Jupiter much faster. Mars has less gravity than the Earth, Venus the same, and Jupiter much more. A whole generation of theoretical meteorologists are testing their models for terrestrial meteorology by applying them to the vast quantity of spacecraft data from these other worlds.[3]

Three little worlds, sired at the same time and raised in the same family, turned out very differently. In the Sun's dysfunctional family, Venus and Mars seem the prodigal children. By studying them, we can understand our Earth better. These planets are tools that help us learn how Earth works, just as we have seen earlier how the Earth's own past is a tool to help us comprehend our present. The atmospheres and surfaces of the other worlds can help explain the patterns of our own, and the absence of advanced life forms on Venus and Mars can underscore how fragile is this gift of life on Earth.

CHAPTER TWELVE

PROBING FOR LIFE
IN OUR GALAXY

Bright star, would I were stedfast as thou art—
Not in lone splendor hung aloft the night
And watching, with eternal lids apart . . .

—JOHN KEATS, "BRIGHT STAR," 1819[1]

Our story of comets and life has taken us far back in time. In this chapter, we explore space in the hope of finding other worlds, probably seeded by comets, where life might have taken hold and arisen. So far, our own solar system has held out little hope. We now move beyond our neighbor worlds to the outer solar system, where conditions are surprisingly different.

EUROPA: A MAGICAL WORLD

Until two spacecraft named *Voyager* visited Europa in 1979, this strange world was an ignored moon of Jupiter, the solar system's largest planet. The second closest of Jupiter's four big satellites, Europa posed for a series of images by the two spacecraft. *Voyager 2*, which got the closer images, revealed a surface of solid ice that was interpreted at the time as being sixty miles thick. With the possible exception of distant Pluto, which may be composed almost entirely of ice, Europa is the only major world in our solar system with such a surface. Europa's surface, the *Voyager* scientists concluded, is very old, perhaps more than 3 billion years.

With its far better images and other data, in 1997, the *Galileo* spacecraft significantly sharpened our view of Europa's ice shield. The shield is far thinner than previously thought, ranging from twelve miles to less than a mile. Moreover, according to *Galileo* science team member Clark Chapman, the lack of true craters suggests that the ice is no older than 20 million years, and that the youngest parts of the ice may be less than a million years old. On the time scale of Europa's life, a million years is like yesterday.

The latest images from the *Galileo* spacecraft show a complex system of breaks in the ice, crisscrossing fractures that seem to be of different ages. The ice almost certainly protects a planet-wide buried ocean that stays warm because of energy from Europa's tides. Just as the Moon pulls on Earth, causing our tides, Jupiter and its inner moon, Io, pull strongly on Europa, causing the moon to stretch and scrunch. Europa's ocean is enormous. More than one hundred miles deep, it cocoons the entire world. If life evolved here, it would be life that could exist in an environment of total water, and in total darkness.

Shortly after *Voyager*'s historical visit to Europa, Arthur C. Clarke wrote *2010*. Based on *Voyager*'s discovery of that moon's mysterious surface of ice, this science fiction landmark probed a philosophical issue: Suppose that Europa's ice protected a liquid subsurface ocean, and that this ocean was warm enough to harbor the beginnings of life. In Clarke's famous story, a distant civilization plants its monolith to teach and to encourage the evolution of intelligent life on Europa.

With the *Galileo* study of Europa now yielding evidence that the icy crust may indeed protect a liquid water ocean, we can ponder the possibility that life has evolved in that ocean. There are life forms that thrive at the greatest depths of Earth's oceans, miles below the surface. No sunlight reaches here, but this oceanic ecosystem thrives near hydrothermal vents from which hot geysers gush, all rich in minerals. Could the same type of life have evolved beneath the icy crust of Europa? Imagine the ecosystems that could have evolved under the protective ice shell! As with many questions posed in

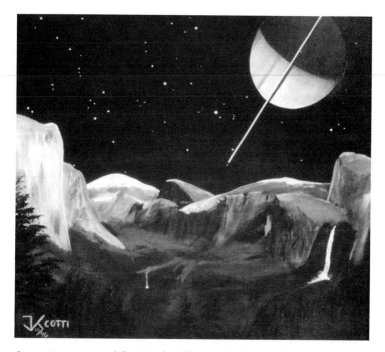

Saturn's moons might, in the distant future, be abodes for life—human life! In this painting, James V. Scotti imagines "what it might be like several hundred years in the future once Mankind has expanded into the far reaches of the Solar System. Humans are likely to terraform various worlds, and as depicted in this work, may like to recreate their favorite Earthly hangouts. Imagine the ease of climbing Half Dome on Enceladus with the gravitational acceleration there being only $1/128$ that on Earth!"

this book, though, I have to leave this one unanswered for now. I do not believe that the answer will come in our generation. In the meantime, however, the question is a marvelous one to ponder.

ON TO SATURN

Worlds encompassed by magnificent rings, Saturn and its nineteen moons are almost 900 million miles from the Sun. These worlds cannot rely on sunlight for warmth. Titan, Saturn's largest moon, is not an abode for life now that its temperature is almost minus 300 degrees Fahrenheit. But Titan's atmosphere contains a rich array of nitrogen and methane. This material reacts with radiation from the Sun—the solar wind—to produce complex organic molecules, a process once believed to herald the start of life on Earth. As we discussed back in chapter 3, Earth's similar primordial atmosphere did not survive its early hot phase, and by the time the planet began to cool, its atmosphere was mostly carbon dioxide. Earth needed an influx of cometary material to regain its hold on life. In the outer solar system, where the primordial worlds never did get very hot, Titan retains its original atmosphere, which is rich in organics. Comets are not required to bring organic compounds to Titan. Too cold for life to begin today, Titan will warm up billions of years from now when the Sun, its hydrogen fuel spent, will swell and engulf Mercury and Venus, and possibly even Earth. In that distant time, life might finally begin on Titan.

COMET PLUTO?

While on our way to the stars, we must pay a visit to Pluto, the ninth major planet of the solar system. Discovered by Clyde Tombaugh in 1930, Pluto is an unusual world that may consist of ices and dust in similar percentages to what we find in comets. Until a few years ago, Pluto appeared to orbit the Sun alone in the dark void of space, accompanied only by its moon, Charon. In 1992, David Jewitt and Jane Luu found the first in a series of small objects, under two hundred miles in diameter, that orbit the Sun beyond the planets Neptune and Pluto. We know very little about these objects, which could be distant large comets populating the long-suspected Kuiper belt, which we discussed in this book's opening chapter. We do know that the outer solar system is turning out to be a busier place than astronomers earlier suspected. There may also be trillions of smaller, Halley-sized comets out there. As king of the Kuiper belt, the ninth planet has an important role to play.

I spoke with Dr. Tombaugh about this possibility a year before his death in 1997. If Pluto really is composed mostly of ices, what would happen if it somehow broke loose from its orbit and hurled into the inner solar system? "Then Pluto," Tombaugh said, his eyes brightening, "would shine as the comet of the ages."

COMETS AND INTELLIGENT LIFE
ON OTHER WORLDS

Could comets have seeded intelligent life forms on worlds in other solar systems? I believe that the answer is yes, but that we may never know for certain. As building blocks, comets are prevalent throughout our galaxy and every other. It is also likely that planets are a common accessory to most stars. If life is indeed a universal imperative, should we not have found other intelligent civilizations willing to talk with us by now?

A LIFE ZONE FOR EVERY STAR

At 93 million miles from the Sun, the Earth is well placed for life's three ingredients to take hold: organic materials, liquid water, and energy. As we have seen in chapter 11, water does not exist in liquid form on Mars or Venus. Thus, the Earth is in the Sun's narrow "life zone," where a life-supporting planet can orbit. A star cooler or weaker than our Sun could also have a life zone closer to it; a hotter star might have a life zone farther away. Of the 200 billion stars in our galaxy, hundreds of thousands might have suitable life zones around them. If the right planet happens to lie within a life zone, then there is a chance that life as we know it could evolve.

We have not explored any of the other solar systems in our own galaxy, and there are trillions of other galaxies in the universe. If we are going to search for life—the end result of comet crashes in the universe—where do we start? An answer comes from simple mathematics, an elegant formula devised years ago by the astronomer Frank Drake. Although it looks complicated, the formula is simple. The letter N on the left is the magic number we often wonder about when we look at the night sky: what is the number of civilizations out there that we can talk with?

$$N = N_* f_p n_e f_l f_i f_c f_L$$

Some might suggest that this equation is a flight of fancy, in which we really know but one number, N_*. That letter represents the number of stars in our galaxy, about 200 billion. All the rest are educated guesses. The phrase f_p is a fraction that limits the search to the fraction of stars that have planets circling them. It is likely that every star has some sort of planetary system, but let's conservatively suggest that only a tenth of the stars in our galaxy have them. That limits our search to 20 billion stars.

OTHER SOLAR SYSTEMS?

Until the mid-1980s, astronomers could only guess that since our Sun had planets, then other stars must have planets, too. In 1984, Brad Smith of the University of Arizona discovered an elongated cloud around the star Beta Pictoris, deep in the southern sky. This cloud could be the early stage of a system of planets. A few years later, Alex Wolszczan and Dale Frail detected three planets around the pulsar unromantically called B1257+12, a designation that refers to its position in the constellation, or group of stars, called Virgo. Pulsars are normally an end result of a star that has become a supernova, and this one spins extremely rapidly, more than six times every thousandth of a second. Even though these planets are approximately the same size as the Earth, the wreck of a star they orbit could not possibly support life—at least not now. In 1995, Michel Mayor and Didier Queloz reported the existence of a Jupiter-sized planet orbiting in the outer atmosphere of the star 51 Pegasi, a sun like our own. Since then, Geoffrey Marcy and R. Paul Butler have reported planets around several stars that are like the Sun, stars named 47 Ursae Majoris, Rho Cancri, Tau Bootis A, Upsilon Andromedae, and Rho Coronae Borealis.[2] Are these planets real? Early in 1997, David Gray, an astronomer at Canada's University of Western Ontario, questioned the existence of the 51 Pegasi planet. This planet is not actually seen but inferred from periodic shifts in the star's position. If Gray is right, 51 Pegasi is pulsating slightly, and these variations, not a new planet, account for the observations.

Because no one has actually seen any of these planets, we cannot say for certain that any of them really exist. But even if they do, none would be even remotely capable of supporting intelligent life. The planet inferred around 51 Pegasi, for example, is the size of Jupiter, and must be incredibly hot since it does hairpin turns around its sun in less than a week.

FINDING PLANETS THROUGH THE DRAKE EQUATION

Someday, I believe, we will find planets that revolve within the life zones of other stars. Returning to the Drake equation ($N = N_* f_p n_e f_l f_i f_c f_L$), let us call the total number of such planets n_e, where the e stands for "ecologically sound for life." If a tenth of the 20 billion suns with planets had this type of world, we'd be left with 2 billion life-zone worlds. Naturally, life will not arise in all these environments. That fraction of worlds where life does start is f_l, and if we use a tenth again, our number is down to 200 million planets. Let's say that on a tenth of those worlds, a fraction called f_i, intelligent life begins; then our number is 20 million. Once such advanced life is present, will it be able to communicate with other intelligent beings? Our humpback whale, while intelligent, does not have a technology to communicate over space. The fraction f_c stands for that fraction of planets where life develops a technological communication medium such as radio. Again, we have no idea what that fraction is. Let's say a tenth again. We're down to 2 million possible planets.

On these 2 million worlds, the last consideration is time. Earth has had life for over 3 billion years, but radio signals have been active for only a century. The fraction f_L represents the length of time that a civilization is active. If that number turns out to be once again a tenth, then the magic number of civilizations we can hope to find and communicate with, in our own galaxy, is still no fewer than 200,000. But in the case of Earth, technology has been around for only $1/40,000$ the length of time that human life has existed here.

If 200,000 is the correct number, we should be receiving signals every night as mighty civilizations across the galaxy crowd the airwaves. But despite years of intensive searching by several groups worldwide under the umbrella organization called SETI, or Search for Extraterrestrial Intelligence, no evidence of intelligent life has yet been found. For those who believe that we are not alone, as I do, this negative result just means that we have not been searching long enough, or in the right way. Or maybe the fraction we used, a tenth, is too high. If we chose a hundredth each time instead, N would actually be less than 1.

From what we've already seen about our early solar system, the rise of life on Earth is actually the result of several happy accidents. One that the Drake formula does not take into account is the presence of a big planet like Jupiter. If Jupiter were not there as the system's vacuum cleaner, the Earth would still be a target for comet hits so often that higher-order life forms would never have evolved here. According to Gene Shoemaker, this constraint severely limits the numbers of worlds where life could arise and progress to intelligence and communication. Shoemaker thought that this limitation was so severe that the magic number N might be just 1. "There might be no one in our galaxy," he concluded, "but us chickens."[3]

On the other hand, someday we may find ourselves members of a galactic club of civilizations communicating with one another. Unlike the fictitious Martian attack of 1938, these advanced civilizations hopefully would be friendly, and the club would be a benign association of civilizations anxious to learn from one another. A discovery someday that we are not alone would be one of history's greatest events. As those civilizations got together, they would probably learn that comets were part of their common heritage.

CHAPTER THIRTEEN

YES, VIRGINIA, COMETS DO HIT PLANETS

In a cosmic sense, the collision of the ninth periodic comet discovered by the team of Carolyn and Gene Shoemaker and David Levy with the planet Jupiter was unremarkable. The history of the solar system, indeed its very genesis, has been marked by countless such events. The cratered surfaces of planetary bodies are a testament to this ubiquitous phenomenon; even the Earth's ephemeral surface records the continued action of this elemental process in impact craters and in the fossil record.

In human terms, on the other hand, the impact of Comet Shoemaker-Levy 9's 20-odd pieces with Jupiter was an unprecedented event of global significance. After a year of planning and preparation, the largest astronomical armada in history focussed on the planet Jupiter in July 1994. News of each successively more astonishing image or spectrum was broadcast with almost instantaneous speed over the world's increasingly sophisticated computer communications network. Astronomers were, for a time, to be found on daily newscasts and on the front pages of newspapers. For a week in July, the world looked up from its normal preoccupations long enough to notice, and to ponder, the awesome beauty of the natural world and the surprising unpredictability of the universe.

—KEITH NOLL, HAROLD WEAVER, AND PAUL FELDMAN, 1996[1]

WHEN I FIRST LOOKED at the stars through a telescope during the summer of 1960, I had no idea what the sky had to offer. My first telescope was a three-inch-diameter reflector that I named Echo. My uncle had purchased it for his children, my cousins, in 1956, for about thirty dollars; when they no longer wanted it, he gave it to me. The summer of 1960 was special for me, for it began with a bicycle accident and a broken arm. A get-well gift of a book about the planets captivated my attention, and by the end of that summer, I had my telescope. On a beautifully clear night in August 1960, my parents and I went outdoors after dinner to try out the new telescope. A telescope without a star chart is definitely not the recommended way to start an interest in the sky, but for me it was enchantment: I had the whole kingdom of stars to look at. I decided to begin with the brightest one. My first glance through the telescope showed the banded planet and four moons that could only be the giant planet Jupiter.

Unknown to me, my parents, or anyone else on Earth that August night, a tiny comet was swinging its way around that distant world. The comet had come a very long way since its birth 4.6 billion years ago in that band we now call the Kuiper belt. For several billion years, the comet moved around the Sun, taking centuries to complete each turn. But at some point, a perturbation took place, and the comet changed to a new heading toward the Sun. Over further aeons, the comet moved in ever-shortening periods around the Sun, until the year 1929. Around that time, according to calculations by Paul Chodas and Don Yeomans of NASA's Jet Propulsion Lab,

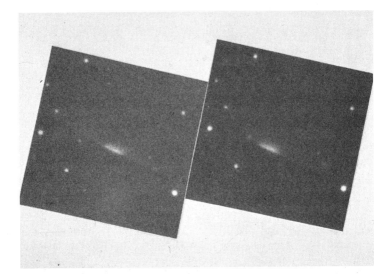

The discovery images of Comet Shoemaker-Levy 9. Photograph by David H. Levy and Gene and Carolyn Shoemaker.

the comet approached Jupiter and was captured by its monstrous gravity.[2] Carolyn Shoemaker was born that June; her future husband, Gene, was a year old that April; and the stock market crashed that October.

On the night of my observing session in 1960, the comet was in its thirteenth orbit of Jupiter. Each orbit around Jupiter lasted about two years, but each was quite different from the one before it. Unlike the Earth, whose orbit around the Sun is almost circular and very stable, the comet's orbit around Jupiter was a long and ever-changing ellipse that brought the comet closer and closer to Jupiter. On July 7, 1992, the comet completed its twenty-fourth orbit by passing a cosmic whisker's distance of thirteen thousand miles from the tops of Jupiter's clouds. As the comet swung past, Jupiter's gravity exerted a tremendous force on the comet, at least two hundred times the force that the Moon uses to cause the tides on Earth. The comet was catastrophically disrupted, and emerged from Jupiter's grasp as a string of tiny fragments.

At the time of disruption, Jupiter was low in the western sky, and difficult to observe from Earth. It must have been quite a sight. After living for billions of years as a single entity encrusted by a dark surface, the comet was

Comet Shoemaker-Levy 9, photographed by James V. Scotti through the Spacewatch telescope shortly after the comet's discovery.

now suddenly split into many pieces, each surrounded by lots of highly reflective dust. But just as no one knew of the comet when I observed Jupiter in 1960, no one was aware of the drama going on far out in space. By the end of 1992, Jupiter was once again a prominent feature in the sky, but the comet, now bright enough to be seen in amateur telescopes, still went unnoticed by anyone.

A THRILLING DISCOVERY

The early part of 1993 was cloudy from southern California's Palomar Observatory, where Gene and Carolyn Shoemaker and I met during January and February to continue our search for asteroids and comets that could pose a threat to Earth (our project is described in chapter 10). Philippe Bendjoya, an astronomer from France, was visiting that month to study our observing strategy. The southwestern United States was suffering from the El Niño weather effect that winter, so clear nights were few and, sadly, far between. The night of March 22, 1993, was one of those few clear nights, and we loaded our telescope with a six-inch-diameter piece of Kodak 4415 film to begin our program. I was just about to shoot the fourth field of sky when I heard some commotion coming from the darkroom downstairs. The first films, Gene discovered in the darkroom, were completely black. During the long period of cloudy weather that preceded this observing run, someone opened our box of precious film and exposed it to light. We had no other fresh film available for that night. As was usual for him, Gene got creative. Realizing that the sheets of film were stored on top of each other, Gene suspected that the top films might have offered some protection to the lower ones. We discarded the

Comet Shoemaker-Levy 9, photographed by Wieslaw Wisniewski through Steward Observatory's ninety-inch-diameter telescope on Kitt Peak. Photo courtesy James V. Scotti.

top five, and found that we could continue with the rest; their edges were exposed to light, but the centers were marginally acceptable. We completed the night with these films.

The next night, March 23, began clear. Armed with a fresh and protected batch of film, we proceeded through our first set of four fields, each photographed twice. But as we began the second set, clouds once more began to interfere. We stopped photographing. Standing outside the dome, we studied the motion and extent of the clouds coming in from the west. They didn't seem to be too thick, so I suggested that we continue for a while. Gene objected, saying that the sky was just not good enough to continue. Fearing that these clouds were the harbinger of a new storm, which would preclude any viewing for several nights, I persisted. Gene's next argument: Each film we use costs about four dollars, and in a good year we might spend eight thousand dollars on film alone. We cannot waste funds on film, he concluded, when the sky was less than clear. It was the financial basis of Gene's reasoning that gave me one final idea. Why not use the remainder of those bad films from the previous night? We had about a dozen left. These would either be used in this way or tossed out, so we would have nothing to waste but our time.

Don Yeomans and Paul Chodas calculated that this could have been Comet Shoemaker-Levy 9's orbital behavior. The comet enters the diagram from the direction of the Sun around 1929, and completes twenty-five orbits of Jupiter, each different from the last, before the impacts in 1994. Image courtesy Donald K. Yeomans and Paul Chodas.

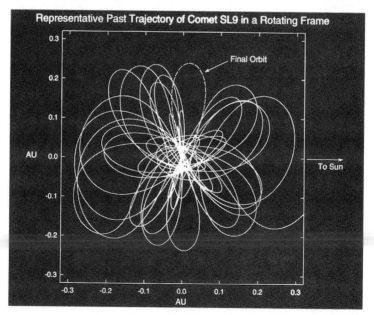

Gene, Carolyn, and I looked at each other, then at the sky. "Let's do it!" Gene said. We resumed observing with the next field on the list, an area of sky highlighted by Jupiter. I proceeded with the next fields but clouds came in again before I could take the crucial second exposures of each area of sky. Without those follow-ups, Carolyn would have no effective way of scanning for any moving asteroids or comets. Miraculously, after almost two hours, a small hole in the clouds approached the Jupiter field. I prepared the film and pointed the telescope. When Jupiter entered the clear area I opened the shutter and barely got eight minutes of exposure before the clouds closed in again.

By 4:00 in the afternoon of March 25, Carolyn had completed her search through all the films we took on the first clear night, and inserted the two Jupiter films. With no new comets on the good films and the weather windy and snowy, we despaired of finding anything exciting on the compromised film we'd used during this observing run. Carolyn slowly moved the films across the stage of her stereomicroscope, then down a bit, then across again. Ten minutes later, just after she crossed the middle of the films, she spotted what looked like a distant galaxy. A few seconds later, she decided to backtrack and look at this object again. This time she stopped. In the eyepieces of her stereomicroscope was a fuzzy line of light, with several tails heading to the north, and two pencil-thin lines off to the east and west. This was no galaxy. "I don't know what this is," she told us, "but it looks like a squashed comet!" Gene stared at this sight, then I looked—it was the strangest thing any of us had seen during all our collective years of comet hunting.

We needed to confirm such a rare find, but the weather outdoors made that task hopeless for us. I called

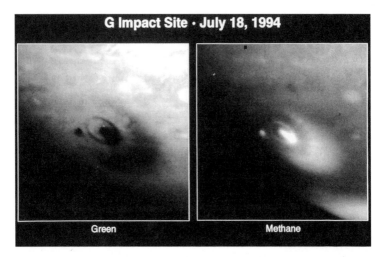

G Impact Site · July 18, 1994

Green   Methane

The crash of Shoemaker-Levy 9's fragment G left a cloud on Jupiter the size of Earth.

my friend and colleague Jim Scotti, who, we hoped, was observing under clearer skies east of us near Tucson. While we awaited his verdict, we met Jean Mueller, an observer on another telescope at Palomar, and she helped us make accurate measurements of the position in the sky of the new comet, data which would be necessary in calculating its orbit. When we returned to our own telescope, I telephoned Jim. So excited he could hardly get a sentence out, Jim described the central fuzzy bar as a train of multiple comet heads, each with its own tail. The east and west lines seemed to be loose dust escaping from the comet train.

AN ARMADA OF TELESCOPES PREPARES FOR COLLISION

In the days and weeks that passed after the discovery of Comet Shoemaker-Levy 9, as the comet was named, astronomers the world over observed it, measured it, and refined our understanding of its orbit. By the middle of April, a picture was emerging: the comet, divided into twenty-one pieces, broke apart in July 1992 because it approached Jupiter too closely. Suspicions were growing by that time that the comet had been in orbit about Jupiter for some time, and that the multiple fragments were still orbiting the planet.

On May 22, 1993, two months after discovery, Brian Marsden issued a remarkable set of International Astronomical Union circulars. "This particular computation," the circular read, "indicates that the comet's minimum distance from the center of Jupiter will be only 0.0003 Astronomical Units (Jupiter's radius being 0.0005 AU) on 1994 July 16.4." In other words, the circular was predicting that sixteen months in the future, the fragments of Comet Shoemaker-Levy 9 would collide with Jupiter.

I have to admit that the full force of that announcement did not hit me until a few weeks later, when I spent several hours at the home of planetary scientist Clark Chapman and his wife, Lynda. As we sat outdoors in the night looking toward Jupiter, Clark explained in simple terms exactly what the collision meant. We learned that evening that the force of the twenty-one comet fragments slamming into Jupiter at the speed of 134,000 miles per hour would be the most dramatic episode ever seen on any planet in all of recorded history. Since we had never seen such a thing, we could not know what to expect, but the sheer energy involved meant that we could get quite a show.

Most astronomers agreed. Within a month, observatories were willingly granting telescope time to study the comet before collision, and during the week of impacts, now predicted to be from July 16 to July 22, 1994.

The summer of 1993 was a scramble to put together the largest telescope armada ever assembled in the history of astronomy to observe a single event. The comet slowed to a relative crawl as it reached the farthest point from Jupiter in its orbit, a distance about as far as Mercury is from the Sun. As the comet began its final journey, instruments and programs for observing the crash were developed, and shared with colleagues in a series of conferences held around the world, from "The Great Crash Bash" in Tucson to a formal three-day observing preparation conference in Baltimore. Meantime, the Hubble Space Telescope took a magnificent picture of all but the last fragment, W, of the comet train. (Shortly after the comet was discovered, the comet fragments were given letter designations by astronomer Zdenek Sekanina.)

On May 20, 1994, the impending collision made the cover of *Time*. As the weeks closed in on the comet's last day, the attention of the world press focused on a planet 477 million miles away. Meantime, the comet was increasing its velocity as it approached its destiny with Jupiter.

A COMET ENTERS HISTORY

The moment of truth came within a few seconds of 4:00 P.M. EDT on Saturday, July 16. As Fragment A was completing its final orbit, its velocity topping thirty-seven miles per second, it tore through Jupiter's upper atmosphere and blew up. A giant plume of material from the comet and the planet shot upward eighteen hundred miles above the tops of Jupiter's clouds. As the fireball settled back down on the planet, a large, very dark spot one-third the size of Earth began to form on the planet.

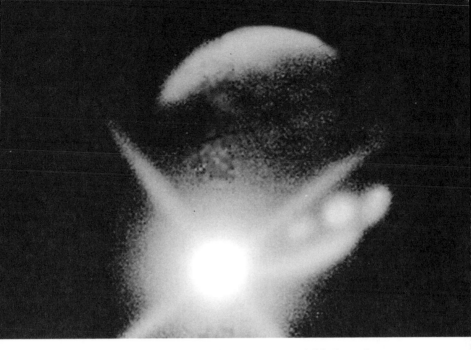

The crash of Shoemaker-Levy 9's fragment K was visible for an hour as material fell back to the planet. The image was taken a few minutes after the impact by Peter McGregor, using the CASPIR instrument attached to the 2.3-meter Australian National University telescope at Siding Spring, Australia.

As hours passed by on that unforgettable night of July 16, E-mail and new reports of observations piled in. "Plumes at CTIO!" was the title of one excited E-mail message from John Spencer, observing from Chile's Cerro Tololo Interamerican Observatory. "Confirmation of very bright plume at 2.3 and 1.7 microns from impact A from CTIO," he wrote, referring to the specific filters he was using. "Brighter than polar caps at 2.36 microns!!"[3] (Because they are so rich in methane, Jupiter's poles are the planet's brightest area when viewed through this filter.) The horde of messages from that evening was a combination of the scientific with the merely elated.

A TRANSFORMED PLANET

With each successive crash, reports flooded in of observers seeing a changed planet through smaller and smaller telescopes. On the night of July 16, Hubble Space Telescope and a few large Earth-based telescopes dominated the reports. As hours and days went by, observations by the thousands were being reported from all over the world. The agony of Jupiter, and the ecstasy of astronomers on Earth, were major news stories. At star parties in cities all over the world, people lined up by scores of telescopes set up to offer a look at a distant planet under siege.

On Monday, July 18, the largest fragment struck Jupiter and set off a fireball brighter, in some filters, than the entire planet. The plume from fragment G lasted for over an hour as matter from the explosion fell back onto the planet's cloud tops. By the time it stopped, Jupiter's southern hemisphere had a dark cloud as big as the Earth. The disruption was so large that even young children saw it in small telescopes their parents had purchased from department stores. Never before had an astronomical event set off such worldwide interest. Virtually every usable telescope in space and on Earth pointed to Jupiter. I recalled that evening with Clark Chapman thirteen months earlier, when he had suggested what could happen. On the night of July 18, he set up his ten-inch-diameter reflector in his backyard and peered at Jupiter. He later told me that waves of awe swept over him when he saw the spots on Jupiter. In a famous note over the Internet, Clark now wrote of his extraordinary observation of a spot created by the impact of a comet fragment that was possibly less than a mile in diameter:

I have just come in from looking at Jupiter with my back yard telescope. . . . Based on my own extensive experience of observing Jupiter when I was younger, and studying historical records of Jupiter observations from the early drawings of Hooke and Cassini through the extensive 19th and 20th century reports of the British Astronomical Association, I would assert: THIS IS THE MOST VISUALLY PROMINENT DISCRETE SPOT EVER OBSERVED ON JUPITER.[4]

THE AFTERMATH

The last major piece of Comet Shoemaker-Levy 9 struck Jupiter on Friday, July 22. It was the only one directly photographed by the *Galileo* spacecraft, then on its way to Jupiter. The actual points of impact were never visible from Earth, since the fragments of the disrupted comet approached Jupiter from the south, then disappeared behind the planet's edge, as seen from Earth, seconds before impact. However, *Galileo* was providentially close enough to Jupiter so that, from its vantage point, the impacting fragments were right in its line of sight.

I love looking at *Galileo*'s image of fragment W, for its implications are as much philosophical as scientific. We do not know when Comet Shoemaker-Levy was born. We suspect that its formation occurred with the rest of the solar system, maybe 4.6 billion years ago. We also think that the comet was captured by Jupiter around 1929, give or take a few years. We do know within an hour or so when the comet broke apart on July 7, 1992. But thanks to the *Galileo* image, we know to the second, 4:06:17 EDT on July 22, when the final fragment of Shoemaker-Levy 9 left the solar system and entered history.

WHAT DOES S-L 9 MEAN FOR US?

"Yes, Virginia, comets really do hit planets." In his closing lecture at the S-L 9 and Jupiter conference held in Baltimore almost a year after the impacts, Gene Shoemaker claimed that was the most important lesson from the now-departed comet.[5] Until Shoemaker-Levy 9, comets and impacts were in the realm of theory. We saw comets with diaphanous tails filling the sky, and we saw craters on the Moon, on Earth, and on other worlds. But until 1994, they could not be connected in a way that would bring their stories to life. With Comet Shoemaker-Levy 9, the link between comets and impacts was clarified sensationally.

The other surprise was the effect the impact of even a small comet or asteroid could have on the Earth. The spots left on Jupiter from each of the largest impacts, those of fragments G, H, K, and L, lasted almost a year, as strong winds spread their effects across the planet's southern hemisphere. If just one of those fragments had hit the Earth, the effect would have been devastating.

The most recent Shoemaker-Levy 9 meeting was held in Paris in the summer of 1996. There, scientists were stunned to learn that an upper-atmosphere haze was still visible over the impact sites as late as June 1995, and that it had spread to more northerly latitudes. Also, two gases created during the impacts, hydrogen cyanide (HCN) and carbon sulfide (CS), were still present in Jupiter's atmosphere, and could stay for years. These findings drive home the idea that an impact over Earth, a far smaller planet than Jupiter, would be dangerous and would leave long-lasting effects.

During the week of impacts, the U.S. House of Representatives, through its Committee on Science and Technology, asked about the feasibility of finding all the comets and asteroids that could present a threat to the Earth. Searching for these bodies, committee members felt, is a priority that deserves the attention of governments around the world.

Comet Shoemaker-Levy 9 brought home the danger we face from cosmic impacts. Of all the threats from Nature that humans face, says David Morrison of NASA's Ames Research Center, "impacts are the one hazard that puts our entire civilization at risk." For the millions of people who watched a battered, injured Jupiter during the summer of 1994, those words have a special and ominous meaning.

CHAPTER FOURTEEN

PRESCRIPTION FOR DOOMSDAY

. . . changed, changed utterly:
A terrible beauty is born.

—WILLIAM BUTLER YEATS, "EASTER 1916," 1916[1]

WHAT WOULD HAPPEN IF a comet exactly like Comet Shoemaker-Levy 9 were to hit the Earth? The story that follows is entirely fiction, but it is based on what we have learned from the Jupiter collision, combined with ideas gleaned from the fossil record. For example, soot in the boundary layer, along with other new evidence, suggests that fires after the dinosaur impact were severe and virtually worldwide. Although this story is fictitious, may Isaac Newton's law of gravitation never conspire to make it come true.

MAY 22, 1999

> This particular computation, [the obscure scientific wording of the circular from the International Astronomical Union read,] indicates that the comet's minimum distance from the center of the Earth will be only 0.00003 Astronomical Units (Earth's radius being 0.00004 AU) on 1999 July 16.4.

The circular's harsh message (see chapter 13 for a text of the circular announcing the 1994 Shoemaker-Levy 9 collision with Jupiter) buried the mention of the collision with Earth in scientific terms. It said only that the comet would come closer to the Earth's center than one Earth radius. The tiny fraction of an astronomical unit, the distance between Earth and Sun, is less than the radius of the Earth. It did not take long for an alert reporter to work through the simple mathematics of the message, and by nightfall, the news was everywhere.

As days rushed by, astronomers all over the world studied the comet with every telescope available, from our own eighteen-inch to the mighty Hubble Space Telescope. The new exact positions allowed the calculation of even better orbits, and by June 15 the chances of a collision of the entire comet train with Earth had increased to 99 percent. Like it or not, a titanic series of collisions was only fourteen months away, and there was nothing anyone could do to stop it.

Everyone had advice. Two days after the May 22 announcement, the U.S. Congress passed a resolution that required NASA to come up with a plan to divert the comet. But we knew that was too late. In order to divert, a rocket armed with a nuclear warhead would need to greet each fragment, and the bomb would have to detonate in such a way that it would not destroy the comet fragment but that its shock wave would gently nudge the comet into a near-miss path.

WHAT WE COULD HAVE DONE

If only someone had found this comet years earlier, the congressional mandate might have worked. But no matter what search program had been in place, astronomers would not likely have discovered this comet before its near-fatal encounter with Jupiter earlier in the decade. The comet was simply too faint before it broke apart that day. Astronomers had known about the possibility of an impact for years. It was to our great disappointment that in the spring of 1999, all the statistics about the possibility of a major impact were now meaningless. For everyone living on Earth, the probability of a collision a little more than a year away was now 100 percent. The tragedy of the coming impact was partly bad timing. In the few years preceding the discovery of the comet, for the first time in

As Comet Hyakutake rushed past Earth in the spring of 1996, its motion among the stars could be tracked easily. In this ninety-minute photo, the comet's head is seen moving differently from the trails of the stars. Photo by David H. Levy.

the history of all the civilizations ever to live on Earth, we could have done something about a body in space headed for Earth. But we did not have enough time.

MAY 30, 1999

President Clinton tried to put the best face on the impend ing disaster. His address to the nation on May 30 was watched by an estimated 90 percent of the country. Clinton announced that the administration's science expert, Vice President Al Gore, would take charge of the task force set up to decide what could be done. Leading the science team would be Bob Goff, a well-known Arizona astronomer who had earned the trust and respect of the comet's discoverers and of many in the comet and impact communities. "Make no mistake," Clinton warned, "our nation, and our planet, face a peril unlike anything we have faced before. It is time for the world community of nations to pull together. May God help us," he invoked, "in the time ahead."

It was quite a surprise that there was not much panic. Actually, it took a few weeks for the severity of the coming impacts to sink in. Virtually all the world's telescopes were following the comets, in an effort to pinpoint the time and place of the hits. The Hubble Space Telescope was quickly able to get a fix on the sizes of the comets as they approached the Earth; the largest piece, fragment G, was measured, at most, to be a mile in diameter.

"This is the first good news we've had," exulted Bob Goff when he heard the revised size estimates. The impact energy would be less than a fifth of what we had been fearing. Also with that announcement, the world's stock markets, which had become sluggish, turned livelier as trading grew more active.

JULY 1, 2000

By July 1, Donald Yeomans and Paul Chodas of the Jet Propulsion Laboratory published their first tentative predictions of when the comets would hit. Traveling at 134,000 miles per hour, fragment A would hit the Earth in the afternoon of July 16, 2000, just off the coast of Los Angeles, and other pieces would hit at the same latitude but at different places on Earth at roughly six-hour intervals over six days.

Also on July 1, the Goff committee proposed its full plan.

1. Launch a reconnaissance mission. First priority, Goff suggested, was to determine precisely what we were dealing with. A small probe could determine how well the comet was put together. Were the pieces solid blocks of rock and ice, or were they flying heaps of rubble? Hoping that they were solid, Goff feared they were rubble piles that would break apart as they got close to Earth, with the result that instead of twenty-one impacts, the impact energy would be spread over hundreds of smaller but even more damaging crashes.

A small spacecraft was quickly readied for launch. It carried a transponder that would be thrown toward the comet's largest fragment, G, as the craft sped by on February 1, 2000. The transponder's purpose: to send a signal back to Earth, allowing the precise measurement of the trail of one of the pieces. Fortunately, the device did find a landing spot somewhere on fragment G, and enabled astronomers to pinpoint G's eventual crash site.

The good news ended there. It was now obvious that the comet fragments were heading straight for Earth's Northern Hemisphere, at latitude 32 degrees. Fragment G, in fact, was aimed directly at the Arizona desert west of Phoenix, the eighth largest city in the nation.

2. Fortify the world's communications systems. The first thing to go in a disaster is the ability to communicate. A program to intensify the land and sea communications lines was immediately enacted. Although it was hoped that some communications satellites would survive the assault by the comets and their associated dust, the emergency plans assumed that they would not.

3. *Prepare emergency shelters, vastly increase firefighting and law enforcement teams, and tell people where to go before impact.* With fourteen months to go, an intense program of preparedness was put together and set up. Millions of citizens volunteered for training duty in law enforcement and firefighting. The strongest buildings in every town and city were designated as safe houses where people could gather. This program became popular: it gave the population the feeling that something was being done despite fears that nothing could be done, and offered a ray of hope that there might be life after July 2000.

4. *Move people away from the impact points.* Goff suggested that in case the pieces were much smaller than suggested, lives might be saved if these cities were evacuated. Los Angeles was directly in the path of what could be a miles-high tidal wave. The announcement of the possible evacuation of the city intensified a sense of panic that until then had not been nearly as severe as expected. The remaining evacuation decisions, it was decided, would be made and carried out quite quickly, after we knew the comet train's final course in the last days before impact. The idea was to follow the Boy Scout motto, and be prepared.

5. *Store a full year's supply of food.* "Impact winter"—a period of darkness and cold temperatures—was a strong likelihood. But this presented tremendous political problems. Who would distribute this food on a worldwide

basis? How would it reach everyone in every nation? Who would be responsible for making sure it was not hoarded or destroyed? Dangerous as it would be, maybe this comet could provide a great incentive and opportunity for the world's nations to pull together to avert a global crisis.

It was not long before the food storage areas in key locations were filling rapidly with canned foods and other supplies, and a fair distribution system was being worked out.

6. Deflect the comets away from Earth altogether. Of all the possibilities for dealing with an incoming comet or asteroid, deflecting it away from us was by far the most desirable, and well-publicized, option. The sad fact was, however, that we had no time to prepare a nuclear bomb and explode it near each fragment. Ideally, a series of such explosions could prod each fragment into a new path that would avoid the Earth. We would do this slowly, a bit at a time, instead of catapulting the objects away all at once, because the smaller blasts needed would be less likely to rip the incoming objects apart, making them even more dangerous.

If we only had the time! If the comet had been discovered ten or twenty years earlier, we might have had the time to meet it and carefully deflect it with a series of explosions. With only a year to impacts, we were fortunate even to have a spacecraft available for the transponder. But Goff's proposed plan hardly mollified a population increasingly terrified and made careless by what was to come. The housing industry collapsed as the population stopped thinking about its future, and as people by the hundreds of thousands stopped making loan payments, the world financial system bordered on collapse. Meanwhile, out in space, fragment A was picking up speed and edging closer to Earth.

JULY 4, 2000

The night before Independence Day, the sky was suddenly rocked by a storm of meteors as the leading wing of dust from the comet began its interaction with Earth. First, there were only a few meteors per hour. By midnight, the numbers had increased to two hundred, then four hundred. People still up before dawn saw the falls increase to forty meteors *per second,* and then an incredible two hundred to three hundred per second. The sky was blazing with meteors, and each night after that the falls continued. For most people, the meteor storm was the first really frightening event, as it brought home the terror of what was to come. As particles hit both weather and communications satellites, the Earth's communication systems began to break down.

JULY 15, 2000: THE LAST PERFECT DAY

Although many people did not believe the predictions of the coming cataclysm, businesses were closed, grocery store shelves were bare, and in Los Angeles, strangely enough, people flocked to the beaches to soak up some sunshine. For once, street-corner philosophers warning people to "Repent! the end is near!" were attracting large crowds. Perhaps it was the general disbelief of the population that prevented riots and looting from breaking out during this time, but for some reason, the world was surprisingly calm.

With clear skies over most of Earth, the view upward was stunning. A bright gibbous Moon shone high in the east, and Jupiter, the planet that had sent the comets on their way two years before, shone innocently in the southwest. The rest of the sky was utterly dominated by three events: the incredible storm of meteors, the crazy motions of the northern lights caused by the interaction of the comet's dust particles with the Earth's magnetic field, and the train of mighty comets. The brightest by far was the first, comet fragment A. Its bright head shone in the west almost as magnificently as the Moon, and its tail stretched across the sky, disappearing over the eastern horizon. The rest of the train stretched out into space in an eerily three-dimensional view: B was at some distance to the northeast of A, C closer to B than B was to A. Even though they were much farther from Earth, fragments G, H, K, and L were almost as bright as A. With long tails stretching parallel to each other, the comet train looked more like a set of celestial Rockettes than a threat: it was a terrifying yet beautiful sight.

JULY 16, 2000

Los Angeles: The day dawned clear, sunny and warm. Overnight, fragment A had brightened so much that it was still visible in the northwest in spite of the Sun and daylight. Every airport on Earth was shut down. Just before one o'clock in the afternoon, fragment A hurtled through the Earth's ionosphere. Its first devastating effect was the utter destruction of the ozone layer. As the fragment blazed into the troposphere, it started to break up into multiple pieces, setting off lightning flashes. It disappeared over the western horizon.

For what seemed like an eternity, but really it was only a minute, all was quiet. Maybe, some dared to hope, that was it. That faith ended as a strange-looking cloud appeared to the west. Suddenly, a crushing gust of wind rushed in, blowing trees first in one direction, then in the other. A deafening roar was followed by a series of earth temblors, measuring 5, then 6, then 8 on the open-ended Richter scale, that toppled buildings, and at the same time came the deafening roar of collision. The wind helped blur the awesome spectacle now fast approaching Los Angeles: a wall of water more than *300 feet high*. It was over in a second, with incredible devastation as untold cubic miles of water flooded the West Coast.

The rest of the nation felt the earthquakes, and within a few minutes, the temperature of the atmosphere started to climb, first to 100 degrees, then 250. After another ten minutes, everyone outdoors was subjected to an atmosphere as hot as an oven set to broiling as particles from the impact crater, created out in the Pacific Ocean, cascaded down on Earth. Within an hour, most of North America was afire as everything that was combustible lit.

By 3:00 P.M., large parts of Earth were covered with clouds of thick smoke from the burning forests and grasslands. But an even more ominous cloud was forming high in the stratosphere—a rapidly thickening pall of thick dust. By 4:00 P.M., the Sun was shining only weakly.

By 7:00 P.M., the obscenely high air temperatures were dropping rapidly. Throughout much of the civilized world, vastly increased firefighting and other emergency services were taxed beyond their limits, but were making some inroads in controlling fires near hospitals and other designated emergency shelters. For those who had been prepared, it appeared as though they might make it—at least past the first impact.

If fragment A were the only projectile hitting Earth, the worst would now be over. But all it did was set the stage. With a sky black with soot from fires and high-altitude dust, no one could see the rest of the comet train closing in on Earth. Early in the morning of Monday, July 17, Israel braced for the assault as the pieces of fragment B began their fall into the Mediterranean Sea. B, it turned out, was different from A. Instead of consisting of a large, solid chunk, B was a cluster of smaller pieces. Within a period of half an hour, a storm of hundreds of pieces blasted through the atmosphere, hitting all areas of the Middle East like shrapnel. Those in the direct line of fire felt the blasts, but no large craters were formed, and there was no great additional contribution to the worldwide pall of dust.

On Monday afternoon, fragment C blasted into the middle of the Pacific Ocean with energy at least equal to fragment A. Once again, the Earth's atmosphere rose to searing temperatures. The Earth's dust cloud was now very thick, and the fires more severe. Despite all the preparation, the sheer magnitude of the disaster was straining the emergency systems. More than at any other time in history, the people needed reassurance, but in most areas, television and radio services were spotty, or completely wiped out.

JULY 18, 2000

This day did not dawn anywhere. Hours earlier, fragment E, larger than the others, crashed with a much greater force into the Atlantic Ocean, and a monstrous five-hundred-foot-high tidal wave virtually wiped out every major city on North America's eastern seaboard, as well as cities on the coasts of South America, Europe, and Africa. But all these events, horrible and catastrophic as they were, were only preludes to the titanic event that was now only minutes away. The severity of the initial falls had been mitigated somewhat by their landing in deep water. Now bolting along at forty miles per second, fragment G, the largest, was headed straight for the Arizona desert only one hundred miles from Phoenix. With sonic booms and a cataclysmic roar, the object collided with such force that it formed a crater fifteen miles wide and one mile deep. Carved out in less than a minute, the crater's unimaginable tons of material rocketed into the atmosphere and crashed back to Earth again. Having limped through the first forty-eight hours, the communications system finally collapsed throughout the United States.

Lurching from disaster to disaster, planet Earth passed through the next days. Fragments H, K, and L struck with the same force as did G. Rain dense with nitric acid was falling, evaporating before it hit the hot ground. Although the remaining impacts were less severe, their combined effect was devastating. At last, on Friday, July 21, the last fragment, W, impacted in a remote mountainous region in western Iran.

The future was really bleak: worldwide fires would rage out of control for several months. When they finally died out, the Earth would be left without crops, with little food, bitter cold, and nothing resembling civilization. The sky was utterly, hopelessly dark—outside was black as pitch. It would be several months before the first faint hint of hazy sun would come to unlock the cloud. Finally, with clearing sky would come a rise in temperature that would last for more than a century. This greenhouse effect would result in melting of the polar caps and flooding of low-lying coastal areas, and increasingly violent seasonal storms like hurricanes.

JULY 24, 2000

On what should have been a bright Monday morning, a tiny group of disheveled people met in the darkness atop the fourteen-thousand-foot Mount Evans, in the Rockies west of Denver. All below was afire; all above was black. Four people held a tallith, a prayer shawl under which stood a rabbi and a young couple. "Do you promise to take this man to be your husband, to love, to honor, and to cherish him?" the rabbi intoned. At that point, the wind grabbed the tallith, blowing it away. The young woman looked at that last small attack from Nature, toward the tallith lost on the mountaintop, then into the blackness, and whispered, "I do." In a small way, at a desolate place, the nightmare veil was lifting at last.

CHAPTER FIFTEEN

AN EPILOGUE

Two roads diverged in a wood, and I—
I took the one less traveled by,
And that has made all the difference.

—ROBERT FROST, "THE ROAD NOT TAKEN," 1916[1]

FORTUNATELY, A COMET LIKE Shoemaker-Levy 9 did not strike the Earth. But some day within the next thousand centuries, and that day could be soon, Earth is likely to have an encounter with a comet or an asteroid the size of one of the S-L 9 fragments. For our story in chapter 14, we did not have sufficient time to try to change the path of the comet so that it would miss the Earth. If a comet were found years before it struck, then we could mount some defense against it.

At our current state of knowledge of all the potential threats to the Earth, the sad fact is that our most likely warning time will be zero. The first hint we will have of trouble would be the blinding flash of light as an asteroid plows through our atmosphere, its shock waves sending off thunderous blasts. If the intruder is a comet, it would probably be discovered by some amateur astronomer using a small backyard telescope, as Comet Hyakutake was less than two months before it passed Earth in 1996.

COMETS AND THE BIG PICTURE

True, there is a chance that a comet will hit, and this chance is what drives a lot of the publicity that comets have received recently. However, that chance is low. The odds of seeing another gorgeous comet cross the sky in the next few years, however, are pretty good. Nothing in the sky quite matches a comet's splendor. On the average, the sky of Earth gets one bright comet each decade. One thing that can't be said is that when you've seen one comet, you've seen them all. Comets are individual,

Comet Levy in 1990, photographed by David H. Levy using the eighteen-inch telescope at Palomar Observatory.

partly because of their composition, and partly because of their varying placements with regard to the Earth. Each of the twenty-eight known visits of Halley's comet, for example, has been unique. In 1910, the comet was very bright, but seventy-six years later, its visit was relatively mediocre. Though the comet's intrinsic performance was virtually identical each visit, in 1986, Earth's viewing point was way at the back of the auditorium. Our view will be better in 2061, but the next time Earth will have a front-row seat will be in 2134.

A COMET A YEAR

Concentrating on the rare great comets does a disservice to the host of fainter visitors that brighten our sky fre-

Comet Levy at its best, photographed by David H. Levy using the eighteen-inch telescope at Palomar Observatory.

quently. Even though these medium-bright comets are seen only through binoculars or telescopes, they are pleasing to watch as they move through the sky with grace, their tails growing and then shrinking over the weeks of their visits. In 1989, Comet Brorsen-Metcalf cruised gracefully through the northern sky, sporting a brilliant green-colored coma and tail. This comet did not cover half the sky, but through a telescope, it was lovely. Comets like Brorsen-Metcalf and Okazaki-Levy-Rudenko in 1989, Swift-Tuttle in 1992, and Tabur in 1996 are examples of the many comets visible through binoculars and that made handsome spectacles in telescopes. We can hope for an average of one such comet a year. Also, each year several fainter comets pass by that are only visible through telescopes. Finally, several more comets, too faint to be followed by a visual look through a telescope, are found each year on CCD frames or photographs.

SHALL WE VISIT A COMET?

Wild 2 is one of these faint comets. Discovered by astrono-mer Paul Wild in 1978, the comet is due for its fourth observed return in 2003. At that time, it may be visited by a spacecraft, designed and built at Caltech's Jet Propulsion Laboratory, called *Stardust*. The craft is slated to sail through Wild 2's coma, collect some cometary particles, and return them to Earth. Early in the new millennium, a European Space Agency mission called *Rosetta* is slated to visit a small comet called Wirtanen and return a sample of its nucleus. "Rosetta" is a good name for a mission set to study a comet, for a comet is really a Rosetta stone that can tell us about ancient times and ancient conditions. It is hard to overestimate the importance of these missions. Through them, we seek confirmation of the idea that comets contain the seeds of life. When *Stardust* and *Rosetta* return to Earth, they may carry our most ancient ances-tors.

CLOSING THOUGHT

Comets are a basic part of the magic of the universe. The farthest comet that we have yet seen, though, comes from the Oort cloud, which is probably no more than a light-year, or some 6 trillion miles, away. Consider that in relation to these words of the English poet Gerard Manley Hopkins: "To know what creation is LOOK AT THE SIZE OF THE WORLD. Speed of light: it would fly six or seven times around the world while the clock ticks once. Yet it takes *thousands of years* to reach us from the Milky Way."[2]

Shining in the constellation of Auriga is a star called

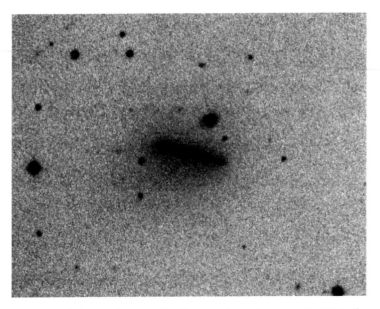

This trailed image of Comet Mueller (catalogued as C/1997 J1) is the comet's discovery photograph taken by Jean Mueller during the course of the Palomar Sky Survey II, using the Oschin Schmidt Telescope.

Elnath. Its light, which we can see on any clear night, left that star about the time Hopkins was writing, some 130 years ago. Elnath is not a particularly distant star, and the Andromeda galaxy, whose light left it some 2 million years ago, is not a very distant galaxy. Both these objects, though faint, can be seen from the backyard of a suburban home. We can hardly imagine the enormity of what Nature offers us through its vast distances and colossal interludes of time. But when a comet swings by the Earth, its flimsy tail stretching across a portion of sky, it brings the universe to us in a manageable dose. The next comet we see will remind us of our birthright, and maybe our ultimate legacy.

NOTES

PREFACE: GHOSTLY APPPARITIONS IN THE NIGHT

1. Leslie C. Peltier, *Starlight Nights* (Cambridge: Sky Publishing Corp., 1980 [1965]), 43.

CHAPTER ONE: FROM DUST TO DUST

1. *The Poetical Works of Gerard Manley Hopkins*, ed. Norman H. MacKenzie (Oxford: Clarendon Press, 1990), 40.
2. 1 Chronicles 21:16.
3. Fred Lawrence Whipple, *The Mystery of Comets* (Washington, D.C.: Smithsonian Institution Press, 1985), 145, 146–47.
4. John Keats, "On First Looking Into Chapman's Homer," in *English Romantic Poetry and Prose,* ed. Russell Noyes (New York: Oxford University Press, 1956), 1129.

CHAPTER TWO: FOUR BILLION YEARS AGO

1. William Shakespeare, *Hamlet,* 2.2. 300–307. *William Shakespeare: The Complete Works,* ed. Peter Alexander (London: Tudor Edition of Collins, 1951, 1964).
2. Thomas Bernatowicz and Robert Walker, "Ancient Stardust in the Laboratory," *Physics Today* 50 (December 1997): 26, 29.
3. *Nature*, September 24, 1997.
4. Wendee Wallach-Levy and Eugene M. Shoemaker, personal communication, July 2, 1997.
5. J. Mayo Greenberg, "What Are Comets Made Of? A Model Based on Interstellar Dust," in *Comets*, ed. L. Wilkening (Tucson: University of Arizona Press, 1982), 131, 157.

CHAPTER THREE: COMETS AND THE ORIGIN OF LIFE

1. Table derived from J. N. Marcus and M. A. Olsen's paper "Biological Implications of Organic Compounds in Comets," in *Comets in the Post-Halley Era,* ed. R. L. Newburn (Dordrecht: Kluwer Academic Publishers, 1991), 449.
2. Armand H. Delsemme, "Nature and History of the Organic Compounds in Comets: An Astrophysical View," in *Comets in the Post-Halley Era,* 416.

3. "Comets and Meteorites: Harbingers of Life on Earth," *Sky and Telescope* 78 (1989): 242.

4. Benton C. Clark, "Primeval Procreative Comet Pond," *Origins of Life and Evolution of the Biosphere* 18 (1988): 209–38.

5. M. Bailey, S. Clube, and W. Napier, *The Origin of Comets* (Oxford: Pergamon Press, 1990), 454–55.

6. T. R. Cech, "RNA as an Enzyme," *Scientific American* 255 (1986): 64–75.

7. One of the best technical reviews of the origin of life is J. N. Marcus and M. A. Olsen's paper "Biological Implications of Organic Compounds in Comets," *Comets in the Post-Halley Era*, 439–62.

8. Amedée Guillemin, *The World of Comets*, trans. and ed. James Glaisher (London: Sampson, Low, 1877).

CHAPTER FOUR: THREE BILLION YEARS AGO

1. Arthur Preston Hankins, *Cole of Spyglass Mountain* (New York: Dodd, Mead and Company, 1923), 302.

2. Mars's position in the sky and position relative to Earth were calculated using Epoch 2000, a graphical-astronomy software program (Meade Instruments Corporation, 1996).

3. Robert Frost, "A Star in a Stone-Boat," in *Robert Frost: Collected Poems, Prose, and Plays*, ed. Richard Poirier and Mark Richardson (New York: Library of America, 1995), 162.

4. D. Levy, McGill University, September 28, 1996.

5. Remarks by the president upon departure, the White House, August 7, 1996.

CHAPTER FIVE: SIXTY-FIVE MILLION YEARS AGO

1. Alfred, Lord Tennyson, *In Memoriam: A Norton Critical Edition*, ed. Robert H. Ross (New York: W. W. Norton and Co., 1973), 78.

2. Walter Alvarez's book *T. Rex and the Crater of Doom* (Princeton: Princeton University Press, 1997) is an accurate and well-written account of the complex story that led to our understanding of what happened in the last day of the Mesozoic era.

3. V. L. Sharpton et al., "New Links Between the Chicxulub Impact Structure and the Cretaceous-Tertiary Boundary," *Nature* 359, no. 6398 (1992): 819–21.

4. "Possible Yucatan Impact Basin," *Sky and Telescope* 63 (1982): 249–50.

5. The story about the Yucatán crater includes details from J. Kelly Beatty's "Killer Crater in the Yucatan?" *Sky and Telescope* 82 (1991): 38–40.

6. The Belize story is well told in several articles by Adriana Ocampo, Kevin Pope, Michael Rampino, Alfred Fischer, and David King, Jr., in *The Planetary Report* 16, no. 4 (July/August 1996).

CHAPTER SIX: COMETS ARE DOUBLE-EDGED SWORDS

1. William Butler Yeats, "The Second Coming," *Collected Poems of W. B. Yeats* (London: Macmillan, 1933, 1977), 210.
2. Peter M. Sheehan and Dale A. Russell, "Faunal Change Following the Cretaceous-Tertiary Impact: Using Paleontological Data to Assess the Hazards of Impacts," in *Hazards Due to Comets and Asteroids*, ed. Tom Gehrels (Tucson: University of Arizona Press, 1994), 882.
3. Ibid., 879.
4. Ibid., 889–90.
5. Roy L. Bishop, *Observer's Handbook* (Toronto: Royal Astronomical Society of Canada, 1998), 25.
6. Michael R. Rampini and Bruce M. Haggerty, "Extraterrestrial Impacts and Mass Extinctions," in *Hazards*, 841.
7. Eugene M. Shoemaker, Ruth Northcott Lecture, Royal Astronomical Society of Canada, Kingston, Ontario, June 30, 1997.

CHAPTER SEVEN: EARTH'S FIRST COLD WAR

1. William Blake, "Auguries of Innocence," in *English Romantic Poetry and Prose*, 222.
2. Grieve and Shoemaker, "The Record of Past Impacts on Earth," in *Hazards*, 425.
3. David Pearson, Laurentian University, personal communication, August 16, 1994.

CHAPTER EIGHT: A TIME FOR COMETS

1. Horace, *Carmina*, I, I, 36, in *The Controversy on the Comets of 1618*, trans. Stillman Drake (Philadelphia: University of Pennsylvania Press, 1960), 19.
2. Seneca, "De Cometis," *Naturales Quaestiones*, ed. Thomas Corcoran (London: W. Heinmann, 1972), VII: 23, 1.
3. This list is adapted from Donald Yeomans's book *Comets: A Chronological History of Observation, Science, Myth, and Folklore* (New York: John Wiley, 1991), 362–424.
4. Yeomans, 43–45. According to the *Oxford English Dictionary*, the swastika symbol originally was supposed to be intended for well-being, fortune, and luck.
5. *Julius Caesar*, 2.2. 29–30.
6. Guillemin, *The World of Comets*.

7. Yeomans, 118.
8. Ibid., 122.
9. Joseph Ashbrook, *The Astronomical Scrapbook* (Cambridge: Sky Publishing Corp., 1984), 46.

CHAPTER NINE: A FIELD OF DREAMS

1. Peltier, 231.
2. Peter Lancaster Brown, *Comets, Meteorites, and Men* (London: Robert Hale, 1973), 128.

CHAPTER TEN: TARGET EARTH

1. J. D. Salinger, *The Catcher in the Rye* (Boston: Little, Brown, 1945, 1951), 173.
2. John Africano, personal communication, November 24, 1997.
3. Brian G. Marsden, personal communication, May 11, 1993.
4. Donald K. Yeomans and Paul W. Chodas, "Predicting Close Approaches of Asteroids and Comets to Earth," in *Hazards*, 246–51.
5. Brian G. Marsden, personal communication, October 31, 1992.
6. This statistic was actually part of "25 Things That Fell from the Sky," in Irving Wallace et al., *The Book of Lists #2* (New York: William Morrow, 1980), 89.

CHAPTER ELEVEN: THREE LITTLE WORLDS

1. Yvonne Pendleton and Jack Farmer, "Life: A Cosmic Imperative?" *Sky and Telescope* 94 (July 1997), 43.
2. Frank Drake, "Carl Sagan, Scientist," *The Planetary Report* 17 (1997): 17.
3. Clark R. Chapman, *Planets of Rock and Ice* (New York: Scribner's, 1982), 98–99.

CHAPTER TWELVE: PROBING FOR LIFE IN OUR GALAXY

1. John Keats, "Bright Star," *English Romantic Poetry and Prose*, 1189.
2. Ken Crosswell, *Planet Quest: The Epic Discovery of Alien Solar Systems* (New York: Free Press, 1997), 251.
3. Eugene M. Shoemaker, personal communication, August 1993.

CHAPTER THIRTEEN: YES, VIRGINIA, COMETS DO HIT PLANETS

1. Keith S. Noll, Harold A. Weaver, and Paul D. Feldman, *The Collision of Comet Shoemaker-Levy 9 and Jupiter* (Cambridge, Eng.: Cambridge University Press, 1996), xiii.
2. "The most likely time of capture," the authors write, "with a probability of 72%, was 1929 plus or minus 9 years." Paul W. Chodas and

Donald K. Yeomans, "The Orbital Motion and Impact Circumstances of Comet Shoemaker-Levy 9," *The Collision of Comet Shoemaker-Levy 9 and Jupiter,* 24.

3. John Spencer et al., Cerro Tololo Interamerican Observatory; SL9 Message Center, University of Maryland, July 16, 1994.
4. Clark R. Chapman, Planetary Science Institute; SL9 Message Center, University of Maryland, July 18, 1994.
5. Eugene M. Shoemaker, closing lecture, Conference on Comet Shoemaker-Levy 9 and Jupiter, Baltimore, Maryland, May 1995.

CHAPTER FOURTEEN: PRESCRIPTION FOR DOOMSDAY

1. William Butler Yeats, "Easter 1916," *Collected Poems,* 205.

CHAPTER FIFTEEN: AN EPILOGUE

1. Robert Frost, "The Road Not Taken," *Collected Poems,* 103.
2. Christopher Devlin, S.J., *The Sermons and Devotional Writings of Gerard Manley Hopkins* (London: Oxford University Press, 1959), 238.

INDEX